B C, BEFORE COMPUTERS

B C, Before Computers

On Information Technology
from Writing to the Age of Digital Data

Stephen Robertson

https://www.openbookpublishers.com

© 2020 Stephen Robertson

ISBN Paperback: 978-1-80064-029-0
ISBN Hardback: 978-1-80064-030-6
ISBN Digital (PDF): 978-1-80064-031-3
ISBN Digital ebook (epub): 978-1-80064-104-4
ISBN Digital ebook (mobi): 978-1-80064-105-1
ISBN Digital (XML): 978-1-80064-106-8
DOI: 10.11647/OBP.0225

Cover image: Katsushika Hokusai (1760-1849), A merchant making up the account. Wikimedia https://commons.wikimedia.org/wiki/File:A_merchant_making_up_the_account.jpg Public Domain.
Cover design: Anna Gatti.

*For Molly and Erin, and Frida and Einar,
and the next generation*

Contents

Acknowledgements

This book started life many years ago as an occasional series of talks to various different audiences, including students at City University, fellow researchers at Microsoft Research Cambridge, and academics at Girton College, Cambridge. Questions and comments from these audiences were immensely helpful.

My wife Georgina, and my son Colin, have both read drafts of this book at different stages and given me a lot of very useful feedback. Michael Williams and another reviewer for Open Book Publishers did the same. Alessandra and the team at Open Book have been unfailingly helpful. Colin also contributed to the design of the cover.

The bibliography at the end lists some books that have given me food for thought. But as indicated there, my primary source for many areas with which I was unfamiliar before researching this book has been Wikipedia. So I would particularly like to thank the huge number of anonymous authors of Wikipedia articles on whose work and insights I have relied.

The mistakes are of course my own.

Stephen Robertson
Norfolk, 2020

Prologue

I was born in 1946. Somewhere around that time was the beginning of a sea change, often proclaimed as a revolution, one which, in the ensuing three-quarters of a century, has transformed our lives in extraordinary ways. Following the pre-war work by such visionaries as Konrad Zuse and Alan Turing, and the inventive necessities of the war-time code-breaking effort at Bletchley Park, the first working computers (in something close to the modern sense of the word) were just being put together in a handful of laboratories in Britain and the United States. Today computers are pervasive—it is hard to identify any aspect of our lives that has not been affected by them.

But computers are only part of it. We can talk about information technology, or more broadly the information and communication technologies, to encompass computers and the digital world that they have made possible, as well the whole of telecommunications, the internet and the web, sound recording and photography and film, broadcasting, and so on. But immediately we have to call into question what I just said about the start of a revolution. The telephone, for example, predates the computer by maybe seventy years (another lifetime); photography by maybe a hundred. So must we then go back another century to look for the start of this revolution? Or perhaps five more centuries, to the invention of printing?

This kind of question is exactly what this book is about. I think it is undeniable that the period I have lived through has seen revolutionary changes in the domain of information technology. But the word *revolution* suggests a complete break, a hiatus, a rupture with the past. It invites us to define when it happened, and to treat this point in time as a discontinuity.

But like all real revolutions, both the start and the origins of this period of huge change are hard to pin down. My contention is that we had to make many other inventions, to devise or learn many ways of thinking about things or of doing things, before the sea change I have lived through

 https://doi.org/10.11647/OBP.0225.14

could come about. What follows is an attempt to pull together into a single story all these necessary precursor technologies, beginning with writing.

This story is not a linear, chronological history. The collection of ideas, of theories, of ways of thinking and ways of doing that have come together under the umbrella of information technology did not start together, either in time or (more importantly) in context. Each strand has its own inception and development; sometimes different strands come together, or one strand splits apart, to follow different historical courses. As a result, I will be jumping about in time, following one strand up to the twentieth century, and putting it aside to go back to the source of another.

Although I have taken the start of the computer age as around 1945, many of the themes that I discuss remained outside the province of the computer or the digital world for much longer. For example, mainstream photography, now absolutely part of the digital world, did not become so until after 1980. In such cases I will follow each theme through into my lifetime, to the point where it is absorbed into or enveloped by this new reality—or perhaps more accurately, until the new ways of thinking and doing expand to include it.

Among the themes that will emerge in a roundabout way is a notion that is key to the modern world—that of *data*. This now all-pervasive idea, which is essentially both the raw material and the product of all computational processes, and encompasses pictures and sounds as well as text and numbers (and a lot of other things), began to emerge explicitly around the start of the twentieth century. It is now hard to think of many aspects of what I will be discussing without this notion in the background. But I invite you to put it aside, at least as far as you can, until Chapter 6.

1. In the beginning...

Ever since the dawn of recorded history, and before, we have been trying to learn how to do things with information.

This is not at all the grand claim it may seem to be. Rather, it is a tautology. We could not begin recorded history until we had ways and means of recording—and recording information is one of the things we have been learning how to do. This is perhaps one of the few necessities of recorded history. We didn't *have* to come down from the trees, or even out of the ocean, before beginning our recorded history—though in fact we did both of those things. We didn't *have* to learn how to plant crops instead of relying on hunting and gathering; we didn't *have* to build towns, invent trade, organise markets and establish trade routes—though probably we did all of those things, and probably they all helped to stimulate the invention of writing. We certainly didn't have to invent the wheel, and indeed it's not clear whether we invented the wheel before or after we learnt how to write. But we did have to learn how to write.

The written message is a specific human invention, just as much as the means to make that message. Once this was invented, information technology had begun to emerge.

Technology

The components of the phrase *information technology* need a little discussion. First, what do I mean by 'technology'?

In today's usage, *technology* is frequently bracketed with *science*, and has come to mean almost exclusively the gadgets and devices that we have invented to allow us to do things—as summed up in the advertising slogan 'the appliance of science'. But this is a very limited view of technology. My (1944 edition) Shorter Oxford Dictionary defines technology as 'a discourse

    https://doi.org/10.11647/OBP.0225.01

or treatise on an art or arts; the scientific study of the practical or industrial arts; practical arts collectively'. It is no accident that this definition contains the word *art(s)* four times and the word *science/scientific* only once. Technology is the art of doing things, of changing the world. We might also think of the word *technique*, concerning ways of doing things, whether in the arts or the sciences.

In this respect, technology is in some sense the opposite of science. Science is about understanding why the universe is as it is; it is about the rules and regulations, and the structures and regularities. Cyril Northcote Parkinson, in *Parkinson's Law*, says

> It is not the business of the botanist to eradicate the weeds. Enough for him if he can tell us just how fast they grow.

The ultimate achievement to which any scientist aspires is the discovery of a law; and a law of nature, just as much as a human law, is about what cannot be done—what possibly imaginable states of the universe are in fact forbidden. There is an old paradox, 'What happens when an irresistible force meets an immovable object?' But Newton has given us some laws, which tell us (among other things) that no force is resistible except by another force, but that any force is resistible by another equal and opposite force; and that no object is immovable.

Technology's view of the world is quite different. For technology, the existence of the universe in its present state is a constant challenge: how do we modify it? How do we mould it to our own ends? How do we avoid these famous scientific laws, or make them work on our behalf, enlist them to our service? Of course, we cannot actually *break* the laws of science (though sometimes technology discovers that the scientists had it wrong, and that what they thought was a law could in fact be evaded). But the ways in which we can make use of them are many and wonderful. The laws of mechanics, including those that govern leverage, are one thing that we learnt about the universe. But when Archimedes said (as a comment on those laws) 'Give me a place to stand, and I shall move the world', he was talking (metaphorically at least) technology, not science.

In short, technology is about changing the universe, about knowing how to do things. Not necessarily on a large scale, of course—in fact some technology is about very small changes. But the *knowing how* is a necessary part.

A tool or device is not technology *per se*; it is only technology insofar as it enables us to do things.

Furthermore, technological change requires choice. Often, technological advances are proclaimed as liberating, as simply expanding our horizons and our opportunities. But as we adopt new ways of doing things through technology, we not only leave behind older ones, we render the older worlds impossible, unattainable (as discussed, for example, in David Rothenberg's *Hand's End*). The huge social change brought about by the availability of the personal automobile, for example, has now spread to almost all corners of the world. Even if the exigencies of climate change fail to force changes in this mode of operating, such changes will perforce come about when the oil runs out. But a return to the pre-car world of 1890 is simply out of the question—we have lost those ways forever.

Information

The word *information* is also a little tricky, and has been used in many ways at different times and by different people. Theses and books have been devoted to the question 'what is information?' and to discussing the consequences of the possible answers (two examples: Luciano Floridi's *Information—A Very Short Introduction*, and Antonio Badia's *The Information Manifold*). However, to go down that route would take me away from my main purpose. In this book I will take a rather naïve view of information. When a speaker speaks and a listener hears and understands; when the speaker's voice is transformed by the telephone handset into electrical signals, and possibly again into radio waves, and something at the other end does the reverse process; when a writer writes and a reader reads and understands; when someone puts data into a database, and later someone else enquires of the database, and gets out these same data in a different form but still understandable—all these are processes involving information.

The only general assumption I shall make is that there are indeed human agents involved at some point (even if I am temporarily concerned only with mechanisms and devices). That is, I shall assume that for some thing to be or carry information, there has to be at least the possibility that the result will at some time reach a human being and be understood. We may think of information as residing somehow in records, and in some sense this book is entirely about records and recorded information—but the human recipient is implicit in everything.

There are certainly notions of information that do not depend on this assumption. However, identifying or understanding some general notion of information that would encompass all these is hard.

There is a well-known theory of information, due in part to Claude Shannon (see his paper of 1948), which uses the idea that 'information is that which reduces uncertainty'. It is possible to read such a definition as requiring no human; however, if we ask "Who or what is experiencing the uncertainty?", it becomes clear that assuming a human (or at least a sentient being) helps with this conception as well.

Shannon himself thought of his theory as having nothing to do with human beings or with meaning. His theory had huge influence in the years that followed, in many different fields, as described by James Gleick in his fascinating book, *The Information*. Gleick's book takes off from many of the same historical starting points as will appear in the pages that follow. To Gleick, Shannon's theory is central to the IT revolution, and encompasses all notions of information. To me, (and also to Badia, cited above) for all its undoubted power and usefulness, this theory fails to address some of the central features of information in the context of human-to-human communication. I shall make no further use of it.

Language

As we have studied animals over the last half-century or so, we have come to realise that many animals have some form of communicative behaviour. We know that bees dance to tell each other about good sources of food, that whales communicate over large stretches of ocean, that chimps learn from each other. Nevertheless, the human behaviours that come under the general heading of *language* are extraordinary in their range and scope.

The evolution and invention of language is one of the great events in human pre-history. I say 'evolution and invention' deliberately. Steven Pinker argues strongly in his book of the same name that 'the language instinct' is exactly that, a basic instinct that has evolved and is part of what makes us human, common to all humanity. In this sense, language in the abstract does not really belong in this history of human invention. However, all those parts of language that are learnt and constructed, all the specifics of actual languages, must be regarded as invention.

The invention of a language is a continuous process. Like most inventions, but even more so than most, languages are not the product of a lone inventor in a garret, but of a social process. Every writer or speaker who uses language inventively or creatively is contributing to the invention of the language, and every writer or speaker who copies or borrows from or imitates a previous speaker or writer is also contributing to the establishment of that invention. Language, in addition to being a basic instinct, is a technology that we use to change the world, by communicating with other people, and the social process by which particular languages develop is one long invention.

We often refer to such a process of social invention as evolution. This is in homage to Darwin's biological theory, but is not a part of it—it is an analogy rather than an application. Probably it is also the case that the biological evolution of humanity in matters relating to language continues to this day—perhaps in obviously physiological features such as voice production, but perhaps also in our language-processing capabilities. That is harder to see happening, however.

It is also hard to know much about the origins of either the social or the biological process, and I will not attempt to go into either. The true starting point of this book, the point from which a genuine technology of information takes off, is the invention of writing.

Writing

The question "What makes humans different from animals?" has been asked many times, and answered in many different ways—including, of course, the answer "They aren't!". Other answers have included intelligence, language, abstract thought and many other things. However, many of these possible answers have been undermined by discoveries in relation to other species. But it may be argued that one characteristic that really does distinguish us from other species is writing. Whether or not you feel the necessity for such a distinguishing feature, writing is certainly a strong candidate.

Writing began, we believe, some time in the fourth millennium BCE—say five-and-a-half millennia ago—in Mesopotamia. A great account of the development of writing is given by Andrew Robinson in *The Story of Writing*. The first purposes of writing were relatively mundane; certainly not the recording of human history. They had to do with commerce and administration—with accounting, recording transactions, listing stock, identifying ownership, and so on. Later writing came to be used to glorify leaders, and to tell stories. These stories (and the people who wrote them) did not distinguish between myth and history. Later still came chronicles and real history, and philosophy and science and religious tracts and laws and administrative rulebooks and poetry and advertising and all the rest.

But the receipts and the tallies and the laundry lists that started it off, however mundane, are central to the history. The kinds of information they represent are the mothers of invention. The people who invented writing felt the need to do several things. They wanted to supplement their own memory, they wanted to organise and impose order on the world, and they wanted to be able inform others of the validity of their own memory and their organisation of the world.

Later I will expand these reasons for writing.

Systems of writing

In order to make this invention of writing work, for any of the purposes mentioned, we need to have a notion of a system of writing. In principle, any mark (on paper or skin or stone or cloth or in clay or whatever) could mean anything we choose it to mean, as in Humpty Dumpty's way with words ('"When *I* use a word ... it means just what I choose it to mean"'—*Through the Looking-Glass*, Lewis Carroll). But that is not a lot of use unless there is some reasonable chance that someone, either the writer at a later date, or someone else to whom the message is directed, will be able to recognise the meaning. So we have to be systematic, at least to some degree, about assigning meaning to marks.

The same problem has occurred long before, in language generally. We have some notion already of the relationship between words and meaning—that is, we can say in some way what a word means, and then construct new sentences out of existing words, whose meaning can be inferred from a knowledge of the words. Of course, this is a gross oversimplification of the notion of meaning, and in any case our present notion of words is rather highly dependent on written language. Consider for example German, where the written form allows some words to be run together to build longer words, or Chinese, which has no word boundaries in its written form. Nevertheless, it is a useful starting place for writing.

At the very least, we might expect our system of writing to tie in with the words of our language, in the sense that the same word is represented in the same way when repeated. This assumes that written language does indeed represent spoken language, and that spoken language is made up of words. It provides us with one of the major ways in which early written languages were constructed—with symbols that may start as stylised pictures representing words.

The system of writing that most clearly illustrates this method is ancient Egyptian—see Figure 1. Small pictures can be seen in many examples of Egyptian writing, and these can sometimes be translated via the words that the pictures represent (though actually 'ancient Egyptian' covers several different writing systems). But even the Cuneiform system, the wedge-shaped marks in wet clay that came out of Mesopotamia, has similar roots.

Figure 1: Ancient Egyptian writing—stela of Senusret III,
Altes Museum Berlin
`https://commons.wikimedia.org/wiki/File:`
`Ancient_egyptian_stela_Senusert_III.JPG`
CC BY-SA 3.0.

In the earliest writing systems, a method based on puns was commonly used. That is, you may want to represent a word about which it is hard to draw a picture (an abstract concept, say). Then one of the methods open to you is to draw a picture of another, more concrete word, which sounds similar, and allow the punning picture to represent the abstract word. This picture now comes to represent the sound rather than the concept that it originally pictured. For example (using modern English words) if I want to represent the word *son* ('my son'), I would find it difficult to represent that meaning directly with a picture. But the language has another word which is pronounced in the same way and represents a very concrete physical object, *sun*. A picture of the sun will do quite well to represent the word 'son', almost certainly unambiguously in the context of a sentence.

This process can be taken one interesting stage further. If the word you want to represent has multiple syllables, it may be hard to find another word that sounds like it. However, you may be able to divide it into smaller parts (shorter words or single syllables or even alphabet-like sounds) and draw a picture to stand for each of the parts. Then you have a representation of the abstract word as a combination of pictures. You have taken a vital step towards a modern system! When each symbol represents a syllable of sound, we have a syllabary. Several such systems were invented, and indeed they still exist in languages such as Chinese and Japanese.

The alphabet

For two thousand years or so, writing systems developed slowly; new systems were invented, and borrowed ideas from each other or introduced new ones, but the changes were not huge. Then, around the late second millennium BCE and the beginning of the first, a huge change took place. The alphabet was invented. In the account that follows, I have oversimplified many things—and borrowed a great deal from John Man's book, *Alphabeta*.

When children in cultures with alphabetic written languages are taught to write, they learn about alphabetic characters and sounds. The idea is that letters represent sounds, and that you can at least to some degree work out what (spoken) word is intended by putting together the sounds of the individual letters of the written word. It seems to be suggested that the natural units of sound are those encapsulated in the letters.

Part of the theory of spoken language is based on the view that the smallest unit of sound that can be distinguished is something like a letter—a phoneme. But this is somewhat misleading, because it is hard to pronounce individual letters. Vowels may be pronounced on their own, but consonants usually need the addition of a vowel before we can actually speak them (hence the way, in English, we vocalise the alphabet as bee, cee, dee, eff etc.). In some sense, the natural units of sound in terms of which spoken language may be understood are not really letter-like; they are much more akin to syllables. Which makes the first stage of developing an alphabet—a syllabary—readily understandable, but lends an air of mystery to the next stage.

So let's step through the process. We have already seen how puns may be used, and how (as a result) a word may be broken into smaller words before it is written down. If we follow that process to its conclusion, we would try to think of an elementary set of single-syllable words, represent them as best we could (by pictograms or whatever), and then construct *all* multi-syllable words as combinations of these elementary words. Modern Chinese illustrates this approach very well. (At this point I am skating over some rather complex notions of the relation between spoken and written language, which certainly come into play with Chinese.)

But a syllabary—a set of symbols to represent every possible spoken syllable—is a clumsy thing. From our vantage point of an alphabetic system, we may think of generating syllables from every possible consonantal sound, followed by every possible vowel sound, followed by every possible consonantal sound. There may easily be thousands of such combinations, therefore thousands of different symbols required, all of which have to be learnt (as any Chinese schoolchild will tell you!).

How could we simplify it? Well, we need a couple of historical accidents. First, we need a language in which consonant sounds are always followed by vowel sounds—so that we can associate each consonant with its following vowel and not the preceding one. Among modern languages, both Japanese and Italian have some of this character. This means that we can get away with an open syllabary or 'abugida' (equivalent to consonant-vowel instead of consonant-vowel-consonant). An open syllabary can be very much smaller than a closed one. Modern Japanese makes use of three different scripts, two of which are essentially open syllabaries. For example, the *hiragana* script has 46 base characters.

As a second historical accident, we need a language in which the vowel sounds do not vary too much. If most consonants, most of the time, are followed by 'ah' sounds, then we may be able to do without the vowels altogether. Modern Arabic is like this—it can be written without the vowels, and still be understood by the reader, because the vowel sounds are sufficiently predictable, and the ambiguities that sometimes arise can generally be resolved easily enough by context. Now all we need is a consonant alphabet or 'abjad'—say 20-30 symbols.

This sequence of events probably took place in the second half of the second millennium BCE, around the eastern end of the Mediterranean and the Horn of Africa. One of the cultures to adopt a consonantal alphabet was that of the Phoenicians, a people who traded throughout the Mediterranean region around the turn of the millennium. The consonantal alphabet was broadcast widely, and its survival was ensured. Note once again that it was the necessities of trade, rather than of literature or philosophy or history or science, that drove this spread.

The final step towards the modern alphabet was an explicit invention, made by the ancient Greeks in the very early first millennium BCE. They observed the Phoenician system and realised just how useful and powerful an alphabetic system of writing could be. Unfortunately their own language was rich in vowel sounds, and would have resisted a purely consonantal solution. So they invented vowels to represent the vowel components of the sounds of language. Well, actually, they borrowed some of the symbols previously used as consonants, but which were not required for their language, and re-assigned them as vowels. And the modern alphabet was born.

Later, of course, the Greeks would invent history and philosophy, and bring science and mathematics and many of the arts to new heights (I exaggerate only slightly!). Beside these, the final step in the invention of the alphabet might seem like small beer. Nevertheless, it is hard to overstress its influence.

As this account suggests, the alphabet was (we believe) invented only once, albeit in several stages. It seems that the syllabary, which is perhaps the more obvious development of the idea of trying to represent the sounds of words, was reinvented more than once. But the alphabet is something altogether more peculiar. Its economy, the fact that we can get away with some 25 symbols to represent *the entirety of our language, including words that have not yet been coined,* is nothing short of astonishing. And its implications are going to reach far into the following 3000 years.

Numbers

Having achieved the astonishing knowledge that we only need a small number of symbols to represent the whole of past and future language, let us put general language aside for a while and think about numbers. Numbers figured strongly in early writing systems, being a very important component of the kinds of information we wanted to represent. And given that we have words for them, we can (in principle) write them down using the same system. However, they do have some peculiar characteristics, which we should perhaps worry about.

For one thing, beyond a certain point we need to be systematic about how we name numbers—we can't simply coin a new name for every new number we come across: there are far too many of them. This applies as much in spoken language as in writing, though whether the development of systematic ways of constructing names for numbers preceded the development of writing is not clear. Secondly, it seems obvious (though again, when it became obvious is not clear) that it makes sense to use the characteristics of numbers to guide us in devising a systematic representation. Thus if we can see a new number as the sum of two numbers for which we already have names, then it might make sense to use the names of the two known numbers to construct a name for the new one. Of course this requires that the idea of addition is already understood, but the early uses of writing suggest that this was the case.

More generally, we would like the representations of numbers (verbal and/or written) to help us with the kinds of operation that we want to do with them. This general principle will take a long time to reach its final fruition—the Arabic number system with which we are familiar today. But in the meantime, early literate civilisations such as the Babylonians and Egyptians developed number systems of some sophistication, and the great mathematicians of classical Greece explored some of the ramifications.

Systems of numbering

Before the alphabet took hold, numbering systems tended to use special symbols. The basic principle, that you have a symbol for each number of a special set, and indicate intermediate numbers (which don't have their own symbols) as sums (additions) of these basic numbers, was established by the Sumerians in Mesopotamia and remained in place until the Arabic system took over. In the Sumerian system, the special numbers were one, six, ten, sixty, six hundred, and so on. Our 60-minute hour is supposed to be a relic of that system. But we are now much more familiar with the Roman system, where the special numbers are one, five, ten, fifty, etc.

By the time of the Romans, the alphabet was established, and they did not need to devise special symbols for their special numbers—they followed the general principle of re-using their existing small set of alphabetic symbols for a new purpose. However, as we shall see later, this is not a true alphabetic solution to the number-representation problem.

The Greeks actually had a variety of number systems. One of their methods had separate symbols for each of the numbers from one to ten, then twenty, thirty, forty, etc. This required 28 symbols to reach what we would now call 900, allowing numbers up to 999. This was a rather profligate use of symbols—they used the letters of their alphabet, but had to borrow a couple of extra ones from someone else's alphabet. Also, 999 was rather early to stop, so they then repeated the alphabet but with a special extra mark on each letter, to get them up to 999,999. But it gave a rather more compact representation of numbers than the Roman one.

A characteristic of all these number systems is that they run out. A point is reached where you have exhausted all the allowable combinations of all the defined symbols, and simply cannot represent the next number (without, that is, defining a new symbol for the purpose). Probably this did not, in general, bother the pragmatic Romans—as engineers and administrators they had the range of numbers that they required, and abstract notions of numbers which they could not represent, but would not need in any case, were of no concern.

But the limitation has some interesting Roman consequences. Consider for example Julius Caesar's books about his military campaigns. His armies or smaller forces are always measured in cohorts and legions, rather than in men. Those were, no doubt, convenient units to use; but it is also the case that Caesar would have had difficulty in expressing the size of his army as a number of men. The Roman system contained both names and symbols up to M (1000), and could therefore represent numbers up to 3999: if we wanted to represent 4000 in the usual Roman system, we would need a symbol for 5000, just as 400 (CD) makes use of the symbol for 500 (D). But a legion was between 3000 and 6000 men, and Caesar normally had several legions under his command. When he referred to the armies against him, he tended to use a mixture of numbers and words, such as 'LX mille'—60 thousand, or 60,000. This is not unlike the modern habit of mixing numbers with the words 'million' or 'billion', but is forced on Caesar. In effect, he has to treat 'one thousand men', in words, as a single unit and then apply the usual numerical system.

However, these limitations on all such number systems most certainly did bother the great classical Greek mathematicians. Archimedes, in the third century BCE, was particularly exercised, and invented his own number system which allowed him to express seriously large numbers (as an illustration, he calculated the number of grains of sand in the universe). But it was still essentially limited by an upper bound on the numbers that could be represented, albeit a very large upper bound.

A *true* alphabetic solution to the number representation problem eluded even the Greeks. It had to wait another millennium or so.

The great Hindu invention

The positional notation that we use today, whereby the same symbol can stand for many different numbers (for example, a "1" can mean one or ten or one hundred, depending on its position) was the revolution we needed. This in turn depended on the zero, as a position marker for an otherwise empty position.

This invention was made by Hindu mathematicians in about the seventh century CE. Though once again, I am summarising with gay abandon a complex and not fully understood process, which may have occurred independently in different places at different times, and/or may have been influenced by earlier ideas—there is a great account in the book by Robert Kaplan, *The Nothing That Is*. The idea that reached the House of Wisdom in

Baghdad, then the centre of the civilised world, was formulated as a general system of both numbering and arithmetic, and re-exported to the rest of the world as the Arabic system—as described in Jim Al-Khalili's *Pathfinders*. One of the masters of this formulation was the Persian mathematician Al-Khuwārizmi, whose name has given us the English word *algorithm*.

The Arabic system provided for numbers what the alphabet had provided for words—a way of representing *any* number, including those for which no-one has yet found a need. The ten decimal digits and the positional notation allow for the representation of any positive whole number. The Arabs also had a precursor to our (relatively modern) decimal point or comma for representing the decimal part of a number—allowing the same rules of arithmetic to apply to non-whole numbers too. It also provided a simple set of rules for arithmetic operations, again applicable to all numbers, which stood us in good stead a millennium later when we began to try to mechanise arithmetic.

Following the original revolutionary invention of writing, these two revolutionary and not at all obvious inventions, the alphabet and the Arabic numbering system, are two of the cornerstones of the developing technology of information. In ways that are quite unimaginable to their progenitors, they will reverberate down the centuries: they will give us ideas, and allow us to think things, that would otherwise have been, quite simply, beyond our ken.

2. Sending messages: the post

Why do we want to write things down? Here are some (not exclusive) reasons:

- in order to organise our thoughts

- in order to remember (remind our future selves)

- in order to communicate with someone else

- in order to communicate with many other people.

The first, organising information, I will discuss later, in Chapter 6. The second, writing as a memory device, I will simply assume. This chapter and the next two are devoted to the idea of sending messages, over space and (usually of necessity) over time. We are concerned with the occasions when the author of the message and the intended recipient(s) are apart, and the message cannot be passed by simply talking across a room.

Messengers

You don't absolutely need to write something down in order to send a message to another person. A human intermediary, who can remember a spoken message, go and find the recipient, and repeat it (exactly or in essence) is of course a perfectly plausible means, which has been used no doubt since spoken language was invented and continues to this day. Many early societies relied heavily on such messengers.

But one of the reasons why writing is so important is exactly that we no longer need to rely on the memory of a single messenger. This method is hardly feasible if the message might have to pass through many intermediaries before it gets to the recipient. If the sender can write the message or

 https://doi.org/10.11647/OBP.0225.02

cause it to be written down, then she can be much more confident that the recipient will receive what she intended, and not some garbled version.

Once you have a system of writing, it is possible to think about systematising the transmission of messages.

The medium

One limitation in this regard is the medium used for writing.

The clay tablets of ancient Mesopotamia were not terribly suitable for carrying around over distances—they were better suited to local record-keeping, individual memory or message transmission over time rather than space. Carved stone is even harder to move around (despite the story of Moses bringing the tablets down from the mountain). So serious letter-writing had to await the invention of a suitably transportable medium.

Over the millennia, several such media have found use. But pride of place in the classical world belongs to papyrus. Made from the dried leaves of the papyrus plant, this medium could be used to construct very substantial messages—whole books were written on papyrus scrolls.

From the time of its invention by the Egyptians, probably in the fourth millennium BCE, the papyrus scroll acquired a huge importance in the affairs of empires. If you want to run an empire extending over a large area, you need effective means of administering it. One requirement is effective communication. In a relatively static hierarchical society such as the Egyptian, where you may have been able to rely on the people in power locally knowing how they were supposed to run their domains, this may not be such a critical requirement. But if you want a dynamic, highly interactive structure, this requires systematic communications. The obvious example here is the Roman Empire.

Many other empires, both earlier and later than the Roman, failed at least in part because they did not have such systematic communications. Of course other things are also necessary, but it is hard to exaggerate the importance of this component. Furthermore, if you are dependent for this on the papyrus plant, control of the papyrus supply becomes a vital factor in the survival of your empire.

Roads

The destination of your message may be just across town, but again, if you have an empire to run, it may be days or weeks away. For a large part of our history, the best way to send anything (goods or letters) across any distance involved boat journeys. But boat journeys are slow and perilous—and they often have to go a long way round. If messengers are to carry your written message at some speed over great distances, they will need roads. Some roads are established simply by people walking them, but your budding empire may need some more reliable and extensive system. Again, the champions here are the Romans.

The Romans are famous for building roads. Straight, well-made roads ran the length and breadth of the Roman empire. For whom were they built? Partly for the soldiers or the administrators: a legion or a governor doing a turn of duty in a remote province would use the established roads where possible, though of course the soldiers at least normally had their main activities in areas not well covered by roads. They may have been built partly for the tradesmen—Rome depended very heavily on trading, and some goods were traded over large distances. But trade was primarily a private concern, and the access that the tradesmen had to roads was a by-product rather than their primary purpose.

But the main reason for the road-building activity of the Romans was for the messengers. The road network, together with the boat routes across and around the Mediterranean, formed the primary communications network of the empire.

In more recent times, for example in the Victorian era, the word 'communications' came to refer just as much to the road and rail networks as to, for example, the postal system. This is no accident. Road and rail, and the shipping lanes, were as much about communicating information as they were about moving people and goods.

The *Cursus*

Efficient empire-wide communication to serve the needs of imperial administration needs to be highly systematic. An official in Rome who wants to instruct another official in one of the far-flung provinces will need to entrust his message to a (human) system, with the confidence that it will reach its destination. Thus was the concept of a postal system born.

Several early empires had postal systems for this purpose; these were not generally accessible to private individuals, but for the use of government only. But, once again, the Roman system introduced by the emperor Augustus was second to none. Called the *Cursus Publicus*, it relied on supplies of messengers running or riding stages and fresh horses at each stage. There were two classes of post—the normal class could be expected to cover 50 miles a day, but urgent letters could go at twice that speed. Its domain was the whole of the Roman empire, and it played a significant role in the success of that institution.

When the Roman empire fell apart, and was replaced by many local administrations, often warring petty kingdoms, both the system of roads and the postal system declined too. The kind of speed with which a Roman official could get a letter to (say) a governor in Gaul was not rivaled again until the end of the eighteenth century.

To the east, some five centuries before Augustus, the Persian emperor Cyrus had initiated a postal system called the *Chapar Khaneh*. Later, after the decline of Rome, during the (relatively) Dark Ages in Europe, a system called the *Barīd* was established in the Islamic world. An account of these systems is given by Adam Silverstein in *Postal Systems in the Pre-Modern Islamic World*.

Postal systems in the ancient world, being primarily organisations for the benefit of the rulers and the government, were closely associated with espionage—one of their main functions was to enable the rulers to discover all they thought they needed to know about what was going on in their domains.

The birth of the modern postal system

The *Cursus* was confined to government business, but in medieval times, some non-government organisations (some universities, for example) were large enough to require their own internal messenger services. The idea of an organisation devoted to providing this service to individuals and other organisations emerged gradually from this need.

The most successful of these private firms, by a long way, was Thurn und Taxis. This started as a private Italian family business, but in the fifteenth century, the family acquired from the Hapsburg emperors a licence—in effect, a state-assigned monopoly—to run all the postal services throughout the Holy Roman Empire. Thurn und Taxis held this monopoly for a little over 300 years. The family were variously ennobled by successive emperors until by the end of the seventeenth century they were princes.

They built a modern and (at its best) highly efficient postal service of a sort we might recognise today. They carried government and private letters, and had an extensive distribution system based, like the *Cursus*, on horse relays with staging posts between the major cities of the empire. It was they who, by the end of the eighteenth century, could rival or beat the kinds of mail delivery speed established by the *Cursus*.

But, again like the *Cursus*, they depended on the authority of the state they served. As stronger national governments developed in Europe, they saw a foreign-run postal service as a threat to their own control over their communications. Countries began to develop their own postal systems. Issues concerning the relation between government and private enterprise, all too familiar today, complicated the development process. On the one hand, some governments preferred a system that was run entirely for their benefit, not serving the public in any way. On the other hand, they were not too keen on any purely private postal service being outside their control. One of the concerns, which again is familiar today, was with security—just think of the horrors that might arise if conspirators were able to communicate freely by letter!

What gradually emerged as the standard approach was to have a government-owned and -run postal monopoly, offering services to the public. The postal charges were often treated by government as a form of taxation, which could be raised to pay for a war or whatever else was required.

A good example of this ambiguous relationship was the experience of William Dockwra in London in the late seventeenth century. He organised a private 'penny post' in London, which quickly became very successful. But its success alarmed the authorities, and they (almost equally quickly) took it over and merged it with the public service.

The Penny Post

Actually, one of the most significant subversive uses of the public postal system arose from the cost of sending a letter. The usual system of payment was for the sender to send the letter without payment, and for the postman to collect the required fee from the recipient on delivery. The fee could be high and quite complex, depending not only on weight but also on the distance travelled and perhaps on the route taken. But it was not hard to work out that simple messages could be coded, for example, by the way the name and address of the recipient was written on the envelope. So when the letter was delivered, the recipient could look at it and then return it to the postman, refusing to pay, on grounds of poverty or whatever—having nevertheless understood the message from the sender.

The obvious solution to this problem, from the point of view of the authorities, was to force prepayment. But it took an enlightened visionary, Rowland Hill (together with another who will reappear later in this book, Charles Babbage), to see that was only part of the solution. Prepayment would actually make the system much more efficient anyway, because delivery would not depend on the postman finding the recipient at home. Hill not only understood this, but also realised that the cost of delivering a letter depended very little on distance, and that a cheaper service would be used very much more widely. When the Penny Post, with pre-payment postage stamps, was introduced in Britain 1840, the effect on the postal service was immediate and far-reaching. It became the universal communications medium, accessible to everyone.

The Universal Postal Union

National postal organisations such as the British Post Office gradually unified and simplified their own internal services, but international mail was a different matter. In order to send an international letter, you would have to know the route and how it was going to be charged by the various carriers involved. Certain national post offices had bilateral agreements with each other, but these might involve a specific fee for each letter. A letter might have to cross several countries in the course of its journey.

All this complication was swept aside in 1874, with the Bern agreement based on Heinrich von Stephan's proposal for a General Postal Union. This laid the foundation for what came to be called the Universal Postal Union. This was a union of national postal services, agreeing to carry each other's international mail to its destination without further charge or accounting on specific items. Initially twenty-two countries joined, but very rapidly it expanded to include practically all postal services throughout the world.

This was a truly revolutionary move, a supra-national agreement to allow a simple system of point-to-point communication across the globe. Anyone could send a letter to anyone else in the world (well, at least to an address, a location). The Universal Postal Union must be regarded as one of the great triumphs of civilisation.

The heyday of post

Universal literacy, with the help of the Penny Post and the Universal Postal Union, ushered in a golden age for postal services. Letter-writing took off as never before. Before radio, before the telephone, long before the arrival of the Internet, the world became a connected place.

From the vantage-point of the twenty-first century, when we have such a variety of ways of communicating, and when the postal service has largely degenerated into a mechanism for delivering purchased goods and spam, it is difficult to imagine the importance of post in the nineteenth and early twentieth centuries. It is also a little difficult to get a grasp of how efficient the service could be. The following letter to the editor of *The Times* of London reveals not only the efficiency (despite the author's protestations to the contrary), but also the importance attached to it:

> *May 25th, 1881*
>
> Sir,—I believe that the inhabitants of London are under the impression that letters posted for delivery within the metropolitan district commonly reach their destination within, at the outside, three hours of the time of postage. I myself, however, have constantly suffered from irregularities in the delivery of letters, and I have now got two instances of neglect which I should really like to have cleared up. I posted a letter in the Gray's Inn post office on Saturday, at half-past 1 o'clock, addressed to a person living close to Westminster Abbey, which was not delivered till next 9 o'clock the same evening, and I posted another letter

in the same post office, addressed to the same place, on Monday morning before 9 o'clock, which was not delivered till past 4 o'clock in the afternoon. Now, sir, why is this? If there is any good reason why letters should not be delivered in less than eight hours after their postage, let the state of the case be understood; but the belief that one can communicate with another person in two or three hours whereas in reality the time required is eight or nine, may be productive of the most disastrous consequences.

I am, Sir, your obedient servant. K.

I would not be surprised if the letterboxes which K used are still there, but if you were to post a letter nowadays, at Gray's Inn at 1.30 p.m. on a Saturday, it would not even be collected from the letterbox before Monday.

The importance of the postal system in the late 19th and early 20th century is indicated by the following statistic: at the start of the First World War, the totality of the Civil Service in Britain was approximately 168,000 people, of whom about 124,000 were employed by the Post Office. During the First World War, the postal service contributed greatly to the public perception of the war, at least for those who were in correspondence with soldiers at the front, which was very far removed from the picture provided by the news media. This sense is vividly conveyed in Vera Britain's book *Testament of Youth*. In a way, despite the inevitable delays of post in wartime, it evokes the kind of feeling of immediacy achieved by television in later conflicts such as Vietnam.

In the 1930s, the British General Post Office produced a wonderful documentary called *Night Mail*. With words by W. H. Auden and music by Benjamin Britten, this short film celebrated a mail-train journey the length of Britain, and at the same time caught the essence of the postal service, as it was seen by the public who used it.

The decline of post

Old media seldom die, but they change. A succession of developments (telegraph, telephone, email and so on) have taken their toll on the concept of a postal service. Paper documents are still important, but for different reasons than those which inspired the letter-writers and -readers of the nineteenth and twentieth centuries. No doubt there are still people in the world who wait on the arrival of the post in the same way that K or Vera Britain did, but this particular manifestation of the global village is surely in decline.

3. Sending messages: electricity

A new medium

The ideal method of sending messages over a distance would not involve the physical transfer of an object at all. The use of bonfire beacons is an old method suitable for a limited number of tasks; slightly more sophisticated is the smoke signal. Both of these have a venerable history. A more recent (eighteenth-century) idea was semaphore, sometimes used for Naval signalling, using hand-held flags or mechanical arms, involving a simple alphabetic code. But major developments in this direction arose from the evolving understanding of electricity. The idea of using electricity for point-to-point communication is almost as old as the serious investigation of electricity as a physical phenomenon. It is certainly older than the notions of using electricity for power, heat or light.

Various systems of signaling using electrical methods were proposed in the early nineteenth century, but the one that had the greatest impact was the system of telegraphy devised by Samuel Morse. This, unlike the earlier proposals, used just one wire, but had a distinct electrical code for each letter of the alphabet. This is the famous Morse Code, consisting of dots and dashes (short and long signals), still occasionally in use today. (At the time of writing this paragraph, one particular brand of mobile phone has, as its default audible signal for the arrival of a text—that is, an SMS—message, the letters SMS in Morse code.)

Morse's electrical system, transmitting codes down a wire, takes us one small step further from visual signals, which is nevertheless a giant leap towards the huge developments of the late twentieth century.

We might also notice how the invention of the alphabet, some three millennia earlier, paved the way. Given that we can construct any message using only the letters of the alphabet (perhaps with a few extra characters such as

     https://doi.org/10.11647/OBP.0225.03

digits and some punctuation), the notion of using a similar small number of codes, which may be manipulated by some physical mechanism, is simple but revolutionary. Now we can transmit *any* message in our language, via writing and the alphabet, using on-off electrical pulses sent along a single wire. It's enough to blow the mind.

The telegraph

As Tom Standage's book *The Victorian Internet* shows us, Morse's telegraph became, around the middle of the nineteenth century, a huge success—not just commercially, but in revolutionising our view of the world in general and communication in particular. Suddenly the speed of physical communication, messengers carrying messages, was no longer the limiting factor in long-distance communication. The achievements of the *Cursus Publicus* and Thurn und Taxis no longer mattered. Provided you had a wire running from A to B, messages could be delivered to all intents and purposes instantly. And wires there were. Networks of telegraph wires spread like wildfire across the developed (and sometimes the less developed) portions of the globe.

But what is most extraordinary about this process is the way in which people suddenly discovered the *necessity* for fast communication, and embraced the medium. Just as the far cheaper and easier penny post was at the same time inviting vast numbers of people to enter the letter-writing age, other groups were discovering the wonders of instant communication. Governments, military authorities, businessmen and news organisations all found it was a medium that they could not do without.

There was never any serious competition between the postal system and the telegraph. The needs for communication expanded to such an extent that both media could simultaneously grow at a prodigious rate. We have not yet reached the heyday of post; the telegraph will in the end turn out to be a rather short-lived medium, because of real competition from the telephone and other media.

Printing telegraph

We are now well into the period of Victorian invention, and many challenges were quickly recognised and taken up by the inventors of the time. Morse telegraphy required a human operator at each end, to make the conversion both ways between the written letters and the dot-dash code. How much easier it would be, people realised, to have machines do these conversions.

Although the eventually successful printing telegraph service, the telex, was a twentieth-century development (actually later than the telephone), there were several nineteenth-century precursors that achieved some degree of success. One of these was due to David Hughes. In the tradition of inventors of the time, he was a polymath; he was eventually honoured as a physicist, and has a Royal Society medal named after him. But in 1855, when he was a professor of music at a college in the United States, he devised a system with a keyboard and a printing wheel. The sender would type out the message letter by letter on marked keys, and the receiving machine would print the message on a sort of ticker tape.

The image that comes to mind from this description is probably the typewriter-like keyboard with which we are now so familiar. I will be talking about the QWERTY keyboard later, but the modern typewriter had not yet been invented in 1855. However, Hughes took his inspiration from the much older keyboard tradition with which he personally was particularly familiar. His keyboard, with alternate black and white keys, looks like nothing so much as that of a piano.

In fact he was not the only, nor even the first, person to consider using something like a piano keyboard for keying alphabetic messages. A slightly earlier device in the vein of printing telegraph was developed by Royal Earl House—his keyboard too was piano-like. In truth, until the invention of the QWERTY keyboard late in the nineteenth century, the piano and its predecessors defined the canonical idea of keyboard control.

Telephony

Even better than writing a message out on a keyboard and then reading a printed version at the other end, would be to speak and hear it. Again, this was a challenge to which the Victorians rose with enthusiasm. In 1876,

Alexander Graham Bell won that particular race by a short head over Elisha Gray. David Hughes was not involved in this race, but he did, within two years of Bell's patent, invent the carbon microphone. The era of the telephone had begun.

But quite quickly, a new dimension was added. The telegraph was a specialist point-to-point messaging system rather like the postal system, with wires strung between offices that acted as gateways for the messages. With telephones, everyone wanted a piece of the action.

The wires had to go to people's homes, and the gateway became the switchboard or exchange, operated by a human being. Directing calls involved connecting one bit of wire to another, via a plugboard. Although manual exchanges continued for a long time, and are familiar to us through films, already in the nineteenth century people were devising automatic exchanges.

The earliest automatic exchanges were of the rotary type. The rotary telephone dial in effect controlled a rotary switch, which moved in synchronisation with the dial. In this system, the number dialled was not held in the exchange (except implicitly in the position of the dials), and not used in any other way. However, already by the 1930s there were exchanges that remembered the dialled digits in a register (like the register in a calculator) and had embedded decision rules about how to route different numbers. This is a form of information processing to which I will return.

Over the course of a century of development of the telephone system, we eventually reached a universal addressing system—a system of numbers defining not only the line on the local exchange, but the exchange itself, then the city or wider area, then the country—so that by the late twentieth-century, a full telephone number represents a single household on the planet. This is comparable to a postal address, somewhat less transparent to a human reader but more amenable to mechanical manipulation.

Radio

By this time, of course, we also had radio: wireless electrical messages. Radio broadcasting will be discussed further below, but it was also used for direct one-to-one communication from very early. Point-to-point radio, radio telephones, international telephone calls routed via satellite, and mobile cellphones, all make use of this medium.

This is a slightly curious development, because radio is naturally a broadcasting medium. That is, a message transmitted by radio can be received by anyone within range and with a suitable receiver. Basically, in order to use it for point-to-point communication, we have to subvert its primary nature. Later, we will see other examples of subverting media to serve other purposes than their nature would suggest.

The technicalities of constructing a temporary link between two telephones for the purpose of making a call (now more like a virtual link than a physical wire) have of course become somewhat more complex, and depend heavily on other late-twentieth-century developments in information technology. The addressing system in the form of telephone numbers has been pushed a little further—now, in the mobile phone age, it designates a unique individual on the globe. Well, that is a slight exaggeration—really it designates a unique phone, but given the present-day spread and use of mobile phones, it's coming close.

Email and text messaging

Perhaps the medium that has provided the closest rival to the postal service is electronic mail. Email systems followed the development of computer networks in the last third of the twentieth century, but really took off with the Internet in the 1990s.

Email is also similar to post in that a one-way message is self-contained—a package with an address on the outside. It does not matter much to the sender or recipient what route it takes; it may go through any number of switches, and some delay at some of the switches is not generally critical. Nevertheless, it took email some time to learn the lessons that the postal services had learnt in the previous century, namely that what was required was a universal addressing system and transparent interfaces between the networks. If you wanted to send a long-distance or international email in the 1970s, you would have had to specify the route to be taken, or at least the main staging-posts along the way. One system used the so-called 'bang notation', leading to an address like this:

```
utzoo!decvax!harpo!eagle!mhtsa!ihnss!ihuxp!grg
```

This means that I want to reach a user called grg, whose mail account lives on a machine called *ihuxp*—but my machine does not know about *ihuxp*. Instead, I tell my mail system to send it to a machine called *utzoo*, which should forward it to *decvax*, which should send it on to *harpo*—with three more intermediate machines before it reaches its destination. In order to send the email, I have to know the route. Furthermore, each staging-post would add another address wrapper around my message, so that even a short message would arrive encased in several layers of headers.

But the Internet and the universal addressing system eventually arrived, and the niche occupied by email in the assembly of communication methods open to us has expanded vastly. For all its similarities to conventional mail, it turns out to have some substantial differences also, and its usage reflects these differences. For example, while it is possible to write the kinds of letters one used to send by post, it is also possible to use email in a much more informal and immediate way—to hold conversations by email that have at least some of the characteristics of spoken conversation.

Another medium that has emerged in the last few years is text messaging. This is a most interesting development, because it has no obvious precursor. As a result, the niche that it has now come to occupy was practically invisible until texting started to become popular (although the informal end of the email spectrum provides some clues). But it shows clearly that despite the huge and obvious advantages of speech, written communication has some distinct advantages of its own. It might be hard for generations not brought up with it to recognise texting as a written form of communication; nevertheless, that is what it is.

A note on electricity

In this chapter, I have regarded electricity purely as a 'medium' for communication. Although we have known tiny bits about electricity for millenia, the serious scientific study of the phenomenon did not begin until around the seventeenth century. But in the nineteenth, we began to discover some of its uses. And our love affair has proceeded at pace. By the end of the nineteenth century, we have made serious inroads into electrical engineering, and have begun to think of it as a resource with many functions. In the twentieth century, it will come to be seen as a vital service to which everyone should have access, with a status almost comparable to the supply of fresh water. Nowadays I have a plethora of electrical devices, and the expectation (even if I am occasionally disappointed) that I can get the electricity needed to run them anywhere in the world, in a standardised form.

And then in the twentieth century our understanding of electricity spawns a monstrous offspring—electronics. Already by 1883 we have photosensors; then the thermionic valve (1904), the flip-flop circuit (the original electronic form of a single-bit memory, 1918), the transistor (1947), integrated circuits (1958), a whole variety of sensors, and so on. In the electronics era, the uses of electricity multiply a thousandfold, leading up to and including the entire digital world.

A full exploration of this aspect of our history would take me too far away from the main themes of this book—though it certainly counts as one of the necessary precursors of the digital age.

The connected world

Now, at the beginning of the third millennium CE, we have a range of methods of communicating with others, which is unparalleled in history. Whether the person we wish to communicate with is in the next office, across the street, across town, the other side of the country, or half way round the world, we have ways to make our messages heard. With a variety of media, at least three global addressing systems, and transparent routing, we are spoilt for choice. In this sense at least, it's a small world.

In the next chapter, we go back again in time, in order to consider the idea of *broadcasting*.

4. Spreading the word

At the beginning of Chapter 2, I talked about writing things down in order to communicate with many other people. We might describe this as *broadcasting*. For much of recorded history, the notion of broadcasting was strongly distinguished from point-to-point messaging. If the originator (a) wants many people to receive the communication, and (b) does not know who all the people might be, the message needs to be thrown out in some sense, like seeds being spread on a field. We will see at the end of this chapter how this distinction might be blurring or even disappearing, and younger people brought up in the era of social media might even find it a little strange or unfamiliar. But its historical importance is huge.

Further, the word *broadcast* is usually associated with radio, and its derivative, television, because of the way the medium of radio is, by its nature, broadcast into the ether. But long before the discovery of radio, a number of technologies were harnessed to the task of spreading messages among many people.

A proclamation read by a crier in a town square is a form of broadcasting (it may be more or less effective in that role, depending on the environment and social structure). Another method that was used extensively in the Middle Ages and for much longer was the pulpit. This mechanism provided not only for broadcasting to a local community, but also allowed a well-organised church to co-ordinate its message across a country or region.

A latter-day form of proclamation, which broadcasts a message to a local community, many of whom might be expected to visit the church or town square, is the poster on a wall. Since the twentieth century, particularly in the west, we have associated posters with commercial advertising, but in some environments they have acquired quite different connotations of public debate. For example, during the cultural revolution in China in the 1960s, major political arguments were conducted through the medium of posters

     https://doi.org/10.11647/OBP.0225.04

on walls.

However, this mode requires a community in which literacy is widespread. During that much longer period of human history in which literacy was a relatively specialised accomplishment, broadcasting via writing took various forms.

The library

The first great mechanism, device, technology that was brought to bear on the problem of broadcasting was the library.

Nowadays we see libraries in various lights—as repositories or archives, as a form of entertainment, as part of the system of education, and so on. Fundamental to these ways of understanding the notion of a library is that libraries, over time, make written information available to many people.

This is, indeed, a major technology. A piece of writing in the pre-computer world can, by its very nature, normally be read by, at most, one person at a time (of course a huge poster may be read by several people at once, but this is the exception). Furthermore, writing on paper is normally in the possession of one person. That person may read it more than once, or may lend it or pass it on to a friend or acquaintance—but this is a very limited form of broadcasting. If broadcasting is seen as desirable, a much more efficient mechanism is required. Given that, in the era we are discussing, we do not yet have the technology for multiple reproduction of a written text, we need to establish a place where people may come and consult different writings, and then to make sure that that place contains all the texts that people might want to consult. Placing a book in a library is broadcasting it—making it available to many people over its potential lifetime, people you do not know.

Libraries have been around for quite a while—in particular, for at least two millennia prior to Gutenberg's invention of movable metal type and the start of the mass printing of books. Indeed, in some sense, libraries were all the more important because there were no mass-produced copies of books. A well-organised archive of clay tablets dating from around 2250 BCE was found at Elba in Syria. We know that there was a library in Assur-bani-pal's palace at Nineveh, in the Tigris-Euphrates basin, when it was sacked in 612 BC—more of that below, as of the Royal Library of the Ptolemys at Alexandria. The House of Wisdom in Baghdad, which I mentioned in the first chapter in connection with the Hindu/Arabic numbering system, was essentially a library and a meeting-place that scholars from all over that world came to visit.

The model that I shall take for my description of the functioning of libraries as broadcasting devices is that of the medieval monasteries in Europe. In the period (the very early Middle Ages, or as they used to be called, the Dark Ages) after the fall of the Roman empire and before the recovery of European civilisation, the monastery system provided the major repository of knowledge and the resources for its spread—the universities came along a little later. But first, a bit about attitudes to libraries.

Burning the library

When Nineveh was overrun (as with Elba more than a millennium earlier), the invaders who sacked the palace also burned down the library it contained. Almost certainly, they had no knowledge that what they were burning was a library—the palace was the seat and symbol of power, but the library was simply part of the palace. As it happens, the burning of the library turned out to be one of the greatest acts of cultural preservation in history. The books in the library used the medium of the time and place—clay tablets. Thousands of these clay tablets were baked hard in the fire, and then buried in ash and sand, and as a result can be seen to this day, in the British Museum. We have, from that event, among many other treasures, the best version of the earliest known written story, the Epic of Gilgamesh. This story was old already, probably a millennium or more, but the survival of the Nineveh version is one of those extraordinarily valuable accidents of history.

By, say, half a millennium or so later, library burning had acquired an altogether different character and meaning. The first emperor of China, in the second century BCE, systematically burnt books because of the subversive ideas they contained—a mode of behaviour repeated many times over the following centuries, including most famously the library at Alexandria. The notion of a library had changed: it had become a repository of knowledge, not a building or an administrative archive, and furthermore the technologies of the time (such as papyrus) were very susceptible to fire.

If you happened to regard knowledge as a bad thing, subversive in some sense—any knowledge or just some of the particular knowledge held in the library—then one recourse open to you was to burn it.

The story goes that the library at Alexandria was subject to this kind of attack, possibly more than once in the early Christian era (a time when subversion of established ideas was not treated lightly). By this time the perpetrators would have known perfectly well what they were burning. Apart from burning down the building, they would (still according to the story) form vigilante patrols to seek out books that had somehow escaped or been rescued from the flames, and burn them as well. This brings to mind Ray Bradbury's *Fahrenheit 451*, about a future world in which the function of 'firemen' is exactly to root out and burn books. This may be fiction, but some of the attitudes it represents have existed in the real world for a couple of millennia.

Actually, the current consensus is that the story of the burning of the Alexandria library is essentially myth. But even as myth, it supports my argument. It was told (certainly from the very early Christian era) as a cautionary tale—burning a library is at the very least an act of cultural vandalism. At its worst, it is an attack on knowledge itself.

The medieval scholar

In Europe, scholarship was associated with religious life. If you wanted to study, you would join a monastic order and seek in that environment the teachers and teachings you would need. And while personal teaching is, of course, as necessary as it has always been, many of the teachings since classical times now reside in books. Your monastery library would contain copies of some of these books.

But every single one has been laboriously copied by hand. While you might eventually hope to write a book yourself, other duties associated with the spread of knowledge may intervene. You may have to undertake arduous journeys to other monasteries to consult books that your library does not hold. And above all, you may have to copy out books by hand, and transport them to other places. For many monks, indeed, copying out other books would be the nearest they would ever get to writing a book.

The business of copying books by hand and carrying them from one library to another was a major occupation of medieval scholarship. At some places and times it acquired an almost industrial flavour. The normal way of copying a book ties up both the original being copied and the monk-scribe for a considerable period, and only produces one extra copy. In one or two monasteries, possessing particularly valuable and sought-after books and many scribes, it was possible to go in for a form of mass-production. A single reader would read the book aloud, and a number of scribes would take it down as dictation. Thus many copies could be produced simultaneously.

So when, in the fifteenth century, Gutenberg's form of printing came along, in some sense the western world was ready and waiting.

Printing and publishing

The ability to reproduce written material exactly, in multiple copies, by mechanical means, was the second great invention to change the face of broadcasting utterly. Once again, I am indebted to John Man's *The Gutenberg Revolution* for this account.

Gutenberg is credited with this invention in Europe, with the proviso that many aspects of printing had previously been invented in the Far East (Gutenberg was probably not aware of this work). From the ninth century, documents were being printed in China, at first with a specially carved wooden printing block for each page, but later with a system of movable type. Individual characters were carved or modelled in clay and stocks of the common characters were built up; rarer characters had to be specially made for a page. In one system in use in the eleventh century, the characters chosen to make up a page were temporarily fixed in resin in a metal frame.

We may note the characteristics of Chinese versus European languages, which might help or hinder this process. Chinese characters are all of the same size, with no breaks between words. Thus the arrangement in the frame is simple—each printed line contains the same number of characters, and the sequence may be broken anywhere for a new line (the same applies if the characters are arranged in columns rather than horizontal lines).

However, the Chinese language suffers one considerable disadvantage compared to European languages: the lack of an alphabet. Chinese has tens of thousands of distinct characters. Even though the number in daily use is somewhat smaller, there is little possibility of building sufficient stocks of characters that every new page can simply be made from stock. And certainly the creation of a mould, from which many new instances of a character can be cast in metal as required, would have made no sense in the Chinese context. Both of these were characteristics of the Gutenberg system.

We can think of printing in economic terms, in a way that may help us to see its revolutionary status. At the core of the industrial revolution is the notion of investing in machinery—to enable the cheap reproduction of goods for which people will pay. The Chinese system of printing involves a lot of investment in the individual printed object—the book or whatever—and might gain a little from a generic investment in printing characters, but not a lot. Gutenberg's system involves a significant prior investment, in the moulds from which the individual characters of type are cast. This makes the typesetting of different books (as well as different pages of a long book) very much cheaper.

This precursor of the industrial revolution is remarkable, not only for being approximately three centuries early, but also for being devoted to the production of information rather than of more material goods. Well, that's an overstatement—books are of course material goods. Nevertheless, their primary value lies in their content rather than their physical nature.

But the mechanical process of printing was only part of the invention. The other part was the system of publishing. The real Gutenberg revolution was to make it possible for the first time for people outside of monasteries or governments to obtain books, build libraries, and take full part in intellectual life and the construction of mankind's fund of knowledge. Publishing joined libraries as a core mechanism for the broadcasting of information.

Publishing

Publishing was not a single datable invention in the way that we might see printing. On the contrary, the idea of publishing started out as a not-very-radical extension of what had been common practice before printing. But the notion has been growing and changing ever since.

When books have to be individually copied, the copies are often (usually) allocated, their destinations predetermined, before they come into existence. In the early days of printing, a book would be prepared for a predefined list of 'subscribers': people who expected, and probably paid upfront, to receive a copy. The idea of printing a large number of copies speculatively, hoping to be able to sell them, emerged only gradually. Also the idea of subscription was transformed, over several centuries, into periodical publications. In the seventeenth century, scientific journals began. If you subscribed to such a publication, you would not know exactly what to expect, but you would have some confidence that it had gone through some selection process before it got into print. Newspapers and other periodical publications eventually followed. Even without subscriptions, the publisher would put out new issues according to some regular schedule, and could reasonably expect many people to buy regularly.

This model has at different times been used for many kinds of publication, not necessarily those we would associate with it today. For example, both the novels of Dickens and the Oxford English Dictionary first appeared in serialised form. Indeed, the model applies at different levels. If you like reading novels, there is some chance that you will try a new novel; if you like Dickens, there is a fair chance that you will try a new Dickens; if you liked the last instalment of *Bleak House*, there is a very high chance that you will try the next.

Far from settling down into some steady state, models of publishing continue to change radically, as we shall see further below.

Cinema

I will later be considering the technologies associated with images and their development over the period of slightly less than two centuries since the invention of photography. However, the role of film as a method of broadcasting belongs here.

Photography itself is no more a natural broadcasting medium than writing. The analogue of the library is the art exhibition or gallery, which has been around for a time and to which photography can contribute. Later, when it becomes feasible to print them in a similar fashion to the printing of text, photographs become part of the publishing world. But film is something different.

In order to see a film, you have to put aside some time and not only reserve that copy of the film for that period, but also have exclusive use of some equipment—including a screen, which means an entire room. The film is equally available to everyone in the room; there is no problem about some people reading faster than others, because the timing is fixed. So it becomes not only feasible but desirable to have a number of people watching at the same time.

The model for bringing people together in this way has existed for millennia already, in the form of the popular show—think, to name but two, of the great playwrights of classical Greece or the Elizabethan theatre. This notion gave birth to the cinema show, one of the dominant broadcasting methods of the early twentieth century. That you can persuade people in large numbers to congregate at fixed times for the purposes of information and entertainment, not knowing exactly what they are going to see and hear, is one of the great social discoveries, repeated over the ages. We can argue that the great populist politicians of the same period, Hitler included, made full and effective use of this discovery, and that the church had previously achieved a similar effect by very different means. But it took the church several centuries, and required a village culture that was receptive to this method of communication.

Although cinema is still around as a method of broadcasting, it is clear that other media such as television not to mention DVDs, the web, and streaming, which do not require people to gather in one place, have challenged and largely overcome its domination. In fact exactly the same might be said of the political rally.

Radio and television

When radio was first developed for communication at the tail end of the nineteenth century, the fact that it was essentially a broadcast medium (that is, if you transmit, anyone within range and with a receiver can hear you) was seen, at least by some people, as a disadvantage. Despite the existence of broadcasting and broadcasting methods for centuries, for many people, the model of communication that came most readily to mind was that of the point-to-point message. The extraordinary success of the telegraph over the previous 50 years no doubt contributed to this view.

Radio was then, and is now, used for point-to-point communication. But the medium of radio had a huge effect on the notion of broadcasting. In one sense, the history of the twentieth century is the history of the development of broadcasting—embracing cinema, radio and television.

Once again, for both radio and television, we require an audience—people who listen or watch speculatively, on the grounds that this particular source has informed, interested or entertained us in the past. This is a little like the traditional notion of a subscriber, though modified for the times. To such an extent has mass communication, broadcasting in all its guises, taken hold in our society that many people spend much of their lives actively or passively open to incoming communication. We expect, all the time, to be aurally or visually entertained.

Copying and printing

Meanwhile, at the turn of the nineteenth century and for about three-quarters of the twentieth, the technologies associated with making multiple copies of documents proceeded apace. In some respects they went in the opposite direction from the printing press and the typewriter—instead of breaking down text into characters, they moved towards the holistic treatment of pages of printable material.

For example, a method of reproducing architects' plans was invented in 1842 and widely used towards the end of the century. This was the *blueprint*, and it has given us a metaphor that has lasted to this day. Original drawings were made on translucent paper, and were reproduced using a simplified photographic process (no camera or lens involved). The wax stencil duplicator was invented in the 1880s—in this case, the original wax stencil was normally prepared on a typewriter although it was also possible to do simple line drawings with a stylus. Given the developments in photography (to be explored further in Chapter 7), it was possible to photograph a document and then make single or multiple prints of it (in the traditional optical-and-chemical photographic method of printing). In the early twentieth century, a photostat machine was developed to make single copies automatically. In this case, the copy was in negative—an original in black type on white paper became a copy in white type on black paper.

But the major development of the twentieth century in document copying was the xerographic process. The main principle was patented in 1942, though the first commercial machine not till 1960. However, it soon made major inroads into the world of business, becoming a ubiquitous presence in offices around the world. The key was another use of electricity—making an electrostatic image on a photosensitive plate, from an original document on paper. The electrostatic image is transferred to paper; black toner particles stick to the charged areas, and are heat-fused onto the paper. No chemical processes were involved, and prints were made on ordinary paper.

Although the Xerox machine printed a single copy at a time, it was very fast, and could easily be used to print multiple copies of an original. Similar techniques were developed for small printing presses, in a process known as offset lithography. Lithography itself as a method of printing, using a prepared flat stone as a printing plate, has a venerable history, having been discovered at the end of the eighteenth century, and used for example by the artist Goya to reproduce pictures, in the nineteenth. But offset litho uses a metal printing plate, normally produced photographically from a paper original.

Both the xerographic process and offset litho have thrived in the digital age. Laser printers use a laser to build an electrostatic image on a printing drum, for printing directly to paper. Similar methods can be used to make a printing plate that can be used for longer print runs. In both cases, the starting point is a computer file, rather than a paper original. Now, of course, in the digital age, virtually all digital objects themselves (in the form of computer files) are indefinitely and accurately copyable, at the touch of a button.

The web

About a century after the development of radio, we discovered a new medium. This is the internet: the vast international network of connected computers. At first glance, the point-to-point wires that make up the internet seem entirely unsuited to broadcasting, in just the same way that radio seems unsuited to point-to-point messages. But just as we have succeeded in subverting the medium of radio to serve many different purposes, including point-to-point messages, we have also subverted the wire-based communication of the internet to devise many new ways of broadcasting.

Of course not all internet connections are wire-based. In fact more and more use is made of radio and other wireless media, to connect computers and other electronic devices, on any scale from centimetres (infrared and Bluetooth) to metres (wi-fi) to thousands of kilometres. But the almost universal arrangement is *first* to subvert the broadcasting medium of wireless to serve a point-to-point function between two computers (which might be your phone and the exchange), and *then* to subvert the multiple point-to-point connections between multiple computers to serve a broadcasting purpose.

The most obvious manifestation of this technology is the World Wide Web. But we have to pay particular attention to the type of system that has emerged as a core component of the web: the search engine. Although the interlinked nature of the web contributes a great deal to its use as a publishing medium, general search engines like Google and Yahoo! and a whole host of specialist search systems have turned out to be critical to its success.

We now seem to be entering a new phase of broadcasting. In the web environment, the reader/listener/viewer/user who used to have to choose a channel and then take a relatively passive role, is suddenly given vastly more control over the communication process. This potential recipient can actively seek out desired information, using a combination of the power of the search engine, the ability to follow links from one page to another, and the ability to recognise what is wanted or needed when it appears on the screen in front of him or her. None of these means is perfect or infallible, but in combination they are very powerful indeed.

The existing publishers or broadcasters, the owners or controllers of the older media, have been having great difficulty in coming to terms with this new medium and its inherent transfer of the locus of control. One battleground in which this conflict is most apparent is that of intellectual property. In the older publishing environments, publishers liked to think they could retain control over the uses of their 'products' even after they had been sold to their customers. This view was already somewhat divorced from reality in the second half of the twentieth century with the arrival of cheap and easy copying facilities (the photocopier as discussed above, the tape recorder, the VCR, the CD/DVD, and finally computer files themselves). But the web has multiplied the opportunities, and therefore the threats to intellectual property, a thousandfold.

Blurring the boundaries

The distinction with which I started Chapter 2, between sending a message to a specific recipient and broadcasting it to anyone who would listen, was a convenient way to discuss a range of different ideas in communication. But it was a somewhat loose distinction, which does not stand up well to detailed examination. The variety of communication methods that we now have at our disposal make the boundary between the two look even more fuzzy.

For example, I can write messages for sending to lists or mailgroups. A mailing list may be something I have created for myself (my siblings, the members of a committee that I run, etc.). Or it can be a list whose membership I know exactly; one whose membership I mostly know but which might include some new members I don't know; a public list or mailgroup where I do not expect to know everyone. It might be a list controlled by one person or one that anybody can join. The knowledge of its existence and/or eligibility for membership may be restricted or widespread. I can post on a blog, which, like depositing my book in a library, opens my message to anyone who finds it in the future. I can tweet a message that might go only to a very few people, or might be picked up by someone with a large following and rebroadcast to a cast of thousands.

If I put a page up on the web, it might be for a particular audience or for general interest. I may link it to some other page that I know is widely accessed, in order to encourage anybody who is interested to visit. I may, in various ways, help the general search engines to find it and to index it in ways I think appropriate, so that a particular but unknown audience can find it easily. Or I may put up a page in order to make it available to a small number of people, those whom in general I would expect to be able to identify. I may, in fact, want to restrict access to those people; this I may try hard to do, by putting serious obstacles in the way of anyone not in that group trying to access this page; or I may try this only in a minor way or not at all.

A similar variety of possibilities arises in the world of printed paper documents. I can easily make any number of copies of anything, from one to a thousand or a million (depending on my resources); I can give them or send them to individuals or to a (paper) mailing list; I can leave copies of a leaflet in some public space, for a limited or a broad audience.

All of these possibilities represent forms of communication somewhere in the no-man's-land between one-to-one communication with another individual and broadcasting.

The connected world—two

The huge variety of communication methods that are now available to us, to which I referred at the end of the last chapter, extend into the realm of broadcasting and into the hinterland just described. Few old media have died: we still have books and journals and newspapers, and radio and television, as well as the web. We have an extraordinary range of devices and methods to help us construct, display, transmit, publish, locate, and access messages of all sorts. Our communication activities, both sending and receiving, can be highly focussed or widely spread or anything in between.

5. More about the alphabet

We have seen how important the alphabet was to many later developments in information technology. Once we have a small alphabet supplying the basic unit from which we can construct text, many other things become much easier or simply possible. Out of the inventions I have discussed so far, the most outstanding examples are movable-type printing and telegraphy.

In this chapter, I will look at four further aspects of the alphabet and alphabetic writing. The first is the way in which we separate our words when we write, mainly with spaces. The second comes directly out of telegraphy, and is the idea of encoding letters or characters with electrical pulses. Eventually, abstracted somewhat from the specific medium, the pulses will become what we now know as bits (the basic units of digital data, as we will see later). The third we have seen briefly already, on the Hughes printing telegraph: it is the keyboard, with each key representing a letter or other character. Finally, I will discuss the abstract notion of a 'character', which both I and the histories I have described have contrived to over-simplify. But first, a small anecdote.

My father was an intellectual, academic, writer, born in 1911. He wrote all his life—books, articles, reviews, letters, poems. His handwriting was all but unreadable; his main method of writing was the mechanical typewriter. He was a competent if sometimes inaccurate typist—I believe he taught himself. Despite being not at all mechanically minded, he came to some accommodation with his typewriters—he even learnt to change the ribbon, which readers of a certain age might just remember as a tricky operation. As a writer, he tended to do a lot of drafting and rewriting. So when, rather late in his life, relatively cheap word processors became available, he eventually acquired one. He was probably in his late seventies or eighty.

As with the typewriter, he became quite good at making the word processor do what he wanted. I don't remember the make, but it had a

 https://doi.org/10.11647/OBP.0225.05

monochrome green text-only screen, a floppy disk, and a more-or-less conventional keyboard. No mouse; no windows: special combinations of control keys would do things like move the cursor around the screen, go to the end of the text, delete whole words, save a document. He would typically open a document on which he had already done a lot of work, go to the end, and start adding to it or editing the previous days' work. But at some point he mentioned to me a small problem that he had. When opening such a document and going to the end, he would often find that the cursor was not actually at the end of the visible text, but considerably further down and to the right. He would have to use the cursor control keys to get back to where he wanted to be.

I worked out that, when deleting words at the end of the existing text (something he did quite often), he would not delete the spaces between them, or the newline characters—so they would accumulate at the end of his text. They were of course invisible—as far as he was concerned, all he had below the text was blank paper. The idea that this apparently empty space was actually part-full of invisible characters is really a very strange one—no wonder he had difficulty recognising it. But I only had to point out to him that he could delete them as well.

Spacing the words

In written English today, as in most writing systems based on alphabets, it is normal to separate the words that we write, by means of spaces—not to mention all the other things that may come between words, such as punctuation marks. This comes so naturally to us that not to do so seems perverse in the extreme. But it was not always the case. Much early writing did not separate the words at all, and even in the Roman period, although some writing would mark the word boundaries in some way, this was not always or consistently applied.

It was not until the sixth and seventh centuries CE that monks in Irish monasteries began to make systematic use of spacing. This was the period when the written culture of the West was largely kept alive in monasteries, in whose libraries books were copied out longhand. A somewhat flip explanation of the introduction of the inter-word space is that the Irish monks were not very good at Latin (the language in which all books were written).

But Paul Saenger, in his book *Space Between Words: the Origins of Silent Reading*, links the practice of word spacing to both the spread of literacy and the practice of reading silently, in one's head. Again, this comes so obviously to us in the universal literacy of the twenty-first century that it is hard to imagine the absence of this practice. But if we go back to classical Greece for a moment, writing was seen in quite a different light. A written text was something a little akin to our present notion of a written musical score: a script for an expert reader to interpret and read out loud to an audience. In such a context, the notion of making it easy for readers simply does not figure.

The spread of the practice of word spacing owes a lot to an English monk. In the late eighth century, Alcuin, a well-known teacher, was invited by King Charlemagne of the Franks to come to his court in Aachen, in order to educate Charlemagne's sons. Among very many contributions to the culture of the court and more widely, Alcuin contributed to the development of a highly legible script (what we might now call a typeface), Carolingian minuscule, and wrote a manual of writing style. It covered many of the things we now take for granted, including punctuation, paragraphs, initial capitals for sentences—as well as spaces between words.

Charlemagne encouraged and presided over a new period of high culture, and eventually became the most powerful man in Europe, reigning as Emperor over a large area. Thus the culture and practices of his court were spread far and wide, and the use of spaces spread to other scripts and languages. We will see below how inter-word spacing comes into the era of telecommunications.

Coding the letters

Although all systems of telegraphy depend on having a small alphabet, the man who saw the connection most clearly was Samuel Morse, together with his collaborator Alfred Vail. His great leap forward was to see that we can take the process one stage further, and work with an 'alphabet' of just two elements: a short and a long electrical pulse, generally referred to as dot and dash. The step is simple, requiring only a small codebook, and makes it very much easier to think about electrical processing of text.

Morse and Vail were very conscious of movable type printing as inspired by some of the same considerations. The Morse-coding scheme involved different length codes for different letters, and they had the inspiration, which is actually the basis for some modern data compression schemes, that it would be most efficient if the commonest letters had short codes. So Vail visited a newspaper printer's workshop to count the stocks of each letter that they kept—because printers know full well exactly what stocks to keep to satisfy most printing requirements.

The idea of encoding the letters in this kind of way has gone through a number of versions since his time. Morse's short and long pulses were designed to allow humans to do the encoding and decoding easily, but we can think of them as any pair of distinguishable states (for example up/down, off/on, black/white). We then need a number of these, spread out in time or space, in groups. A code book lists each distinct group with the object it represents—for example, the Morse code book says that the letter A is represented by the group *dash-dot*, and D by *dash-dot-dot*. In this case, the things we want to represent are the letters of our usual alphabet.

If we think of this pair of elements as itself an 'alphabet'—in the abstract, rather than in a specific physical form—what we have is the modern concept of a 'bit', a binary digit. Usually nowadays we think of the two states as 0 and 1. So the letter A in Morse is 01, and D is 100. (Morse code uses different numbers of bits for different letters, but most schemes allocate codes in fixed-size groups.)

Actually, Morse was not the first to use such a binary coding scheme for letters. The familiar Braille system of embossed dots on paper (invented by Louis Braille), designed to allow the blind to read with their fingers, predates Morse code by a decade or so. The dots are in groups of six—that is, the group is a rectangular array with six positions, in each of which the dot is either present or absent. The codebook specifies which dots are actually present for each letter.

The later Baudot code (invented by Émile Baudot) used for telex is a fixed-length, 5-bit code. A small calculation will show you that this gives 32 different combinations—enough for the 26 letters of the Latin alphabet, though not for upper and lower case. Actually this is not quite enough, even if we do not care about case—it does not allow for any punctuation marks or the digits (Morse has codes for the ten digits and one or two punctuation marks). For this reason the Baudot scheme includes a shift code—a little like the shift key on a keyboard, or more precisely a caps-lock key—which doubles up the meanings of the remaining codes. Braille uses a similar method for extending the range of characters represented.

Coding for the modern era

As we entered the computer age in the 1960s, new coding systems were defined. In fact there were two main rival schemes, EBCDIC (pronounced *ebsidik*, for IBM machines) and ASCII (pronounced *askey*, for all other computer manufacturers). I will leave EBCDIC aside, but ASCII is worth some discussion. The American Standard Code for Information Interchange is a more ambitious system than Baudot, and was used for multiple purposes in the transmission and storage of data in the early computer age, and in fact is still in use. It is a seven-bit code, allowing a total of 132 different combinations. These include the 26 letters, in both upper and lower case (making 52), the ten digits (62), a significant number of punctuation marks and special symbols (96), and 32 codes reserved for control purposes. Minor variations on this system were defined for various European languages with features not seen in English, e.g. accented characters. More systematic variation is provided for by the scheme known as ANSI, which started from ASCII, but has different code pages for different languages. Each code page provides a complete coding of a set of characters for a language—but the computer must 'know' which code page is in use to interpret ANSI correctly.

The coding scheme known as Unicode, which is currently becoming the standard for many purposes, is a much larger set. It includes not only all the characters for other alphabets than the Latin one, e.g. Greek, Russian and Arabic, but also characters for non-alphabetic languages, e.g. Japanese, Chinese. This is a fascinating development: the *idea* of a coding of characters could only have developed in the context of a small alphabet; but given the idea, it now becomes possible to apply it to much larger character sets. Unicode in its original full form requires 16 or 32 bits per character, but there are alternative encodings for the same scheme, which allows the old ASCII character set to be represented as it traditionally was, in eight bits. (Yes, I know I said seven. ASCII is a seven-bit code, but since most computers operate with multiples of eight bits, ASCII is usually embedded in eight bits.)

Although there remain some languages and scripts in the world that have not yet been incorporated into the scheme, nevertheless we seem (in the early twenty-first century) to be approaching the state where any text character in any language can be represented by means of a standard binary code. This is a remarkable achievement.

The last alphabet

The bit—the binary digit, a character from a two-letter 'alphabet'—might be seen as the final stage of a process that began when we started inventing systems for writing, something like five-and-a-half millennia ago. At the start of the third millennium CE, we realise that we can represent *any* record by means of bits. Not just language, but also, as we shall see in later chapters, numbers, images, sounds, moving pictures, and so on. The universal alphabet consists of just two symbols, a zero and a one.

But it is not just a matter of representation. We have already seen how the alphabet has helped us towards new ways of *doing* things with information. Printing, and thus the publishing revolution that followed it; and Morse code, and the telegraphic revolution that followed that: each of these would have been inconceivable if we had not invented the alphabet in the first place. Now, in the new revolution, the biggest changes have come about in the ways of *processing* information: of systems and methods and mechanisms that operate on information rather in the way that a loom operates on the raw material of thread to produce something quite different, cloth. Sending messages was just one such operation; but the possibilities are almost limitless.

The strange story of the keyboard

I am writing this text by means of a device that has become so common that it passes almost without notice: a QWERTY keyboard. This is something else that the alphabet made possible. Although my keyboard has rather more than 26 keys, its existence depends on the small number of possible characters—a keyboard with a key for each of the tens of thousands of distinct Chinese characters is quite inconceivable.

We have had keyboards for musical instruments for centuries. But the idea of associating keys with letters of the alphabet (or with numbers, come to that) has been around since the mid-nineteenth century at least. As we have already seen, Hughes used a piano-like keyboard for his telex-like machine. Various attempts were made to develop typewriters from earlier in that century.

But the development of an effective and useable typewriter had to wait until a little later. The primary inventor was Christopher Sholes; over a period from the 1860s until the 1890s, he and Remington, the company he worked with and eventually sold out to, pushed the typewriter from the status of one of those fascinating but impractical Victorian inventions to that of a common business accoutrement. To do this, Sholes had to solve a number of tricky mechanical problems. The design and layout of the keyboard he produced, as a result of confronting these mechanical problems, is with us to this day. If Sholes were to walk into your twenty-first-century office, one of the very few things he would recognise would be the QWERTY sequence on your computer keyboard. That this is so is even more extraordinary than you could possibly have imagined.

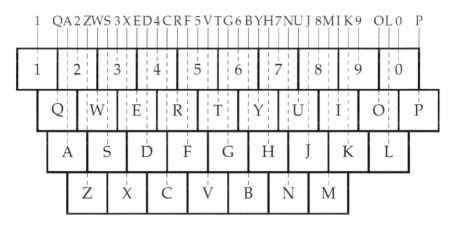

Figure 2: Basic QWERTY keyboard (with the interleaved bars of a traditional typewriter). Diagram: the author.

It's not just the sequence of letters we are talking about here. Look at Figure 2 (if you have a keyboard close to you, compare it to the diagram). In particular, look at the way that successive rows of keys are offset from each other. Notice that the ZXCVB line is offset from the ASDFG line by half a key width; in other words, Z is half way between A and S. Also the QWERTY line has a half-key offset from the numeral line. But the offset between the QWERTY line and the ASDFG line is—what?—one-quarter of a key? But why on earth?

No, it is most certainly not any ergonomic or ease-of-use reason; in fact it is quite hard to learn to use. The reason is purely mechanical. Imagine that each key perches on the end of a metal bar, which comes out of the back of the machine (where the paper would have been). These bars have to be straight and parallel—or their movement would not be true when pressed—and cannot be allowed to interfere with each other. So they have to be carefully interleaved. The Q bar goes between the 1 bar and the 2 bar, and W between 2 and 3, that's easy. But now the A bar has to go between the Q bar and the 2 bar, S between W and 3; and Z between 2 and W, and so on. Now you see why it had to be so.

It is just possible that you have a keyboard that does not follow this off-set convention. Some PDAs and other small machines, and tablets with on-screen keyboards, have the QWERTY sequence but use either no offset at all or a universal half-key offset. But real keyboards invariably use the Sholes offsets. Even if you are (say) French, and have one of those keyboards where the letter sequence is AZERTY or some other variation on Sholes, you will still have those offsets. Some keyboards are split into two parts, for ergonomic reasons associated with the way you place your hands; but they *still* use the Sholes offsets on each half.

But, you may argue, my keyboard no longer has those metal bars: indeed, you have to be of a certain age even to remember their existence. Nowadays, each key operates its own microswitch, and they could be arranged in any way we choose. So why do we persist in using these offsets? Well, this is part of the story.

Keyboard wars

In the 1880s and '90s, several rival typewriter companies were formed, and a number of different keyboard arrangements were in use. Another characteristic of the Sholes keyboard is that when they introduced lower as well as upper case, they did this by means of the familiar shift key—which has now, of course, been joined by a few imitators, like the CTRL control and ALT alternate keys. But at least one of the rival companies just added more keys, so that upper and lower case letters were on separate keys.

As typewriting became more common, schools to train typists were set up, and various systems of fingering were devised for the different keyboards to help typists work faster—the earliest typists were almost certainly one- or two-finger typists. Claims and counter-claims were made about the relative speeds of these different combinations. And pretty soon, they became competitions.

An 8-finger method of typing was devised by Margaret Longley, who ran such a school, in the early 1880s. She applied this method to different makes of typewriter—but as applied to the Sholes keyboard, it is similar to the fingering taught today. Frank McGurrin, a court stenographer, used it with great skill on an early Remington. Another student and later principal of the Longley school, Louis Traub, used a similar fingering on Caligraph machine with a six-row keyboard.

The first competition, in 1888, pitched Traub against McGurrin. But McGurrin had a card up his sleeve. The trick was that he had discovered that he could memorise the keyboard layout, not looking at the keyboard while typing, but at the paper (he could also type blindfolded). He invented what we now know as touch-typing.

This turned out to be the ace. McGurrin thoroughly beat Taub, who shortly afterwards switched to a Remington. McGurrin went on to win many more competitions; and the keyboard never looked back. Gradually, the rival companies adopted the Sholes layout. Having a single system, a standard keyboard layout and method of typing, was a big advantage from the labour point of view. There is a delightful account of this event in Stephen Jay Gould's essay *The Panda's Thumb of Technology*, published in the collection *Bully for Brontosaurus*.

In the twentieth century, it was common to denigrate the Sholes keyboard, and to claim that it is very inefficient and unergonomic for the typist (even that it was *designed* to slow the typist down, which is not actually the case). A rival system was designed on ergonomic grounds, the Dvorak keyboard, which has the same basic structure as Sholes but a very different arrangement of the letters. In a series of experiments, it was demonstrated that Dvorak was easier to learn and faster to type on than Sholes. However, the Sholes keyboard was so well established by then that it proved impossible to dislodge. In fact, the experiments (rather like the 1888 competition) were somewhat suspect as scientific evidence; probably the differences are not very large. Besides, Dvorak did nothing at all about the offsets, which are certainly one of the sources of ergonomic problems with the Sholes keyboard.

In the second half of the twentieth century, we saw the development of (successively, inter alia) the IBM golf-ball typewriter, the word processor, the PC, the laptop. With each of these developments, we could, in principle, have abandoned Sholes and devised something that might have been better. But this is not the way things work: technologies have to co-exist; people have to switch between them; people have to maximise the benefit they get from the investment they have put into learning something. If you are an experienced typist, your fingers remember not only the locations of the letters, but also the offsets. Even moving ASDFG a quarter-key to the right, so that all the key offsets are half-keys, would confuse you.

I once pointed out the offsets to a man with touch-typing skills who had managed to transfer them to one of those tiny PDA keyboards, about ten centimetres wide. The designers of this keyboard had retained the QWERTY layout, but (obviously not expecting anyone actually to touch-type on it) had made all the offsets a half-key. His instant response was "I *knew* there was *something* wrong with it!".

All these things conspired to ensure the persistence of almost every aspect of the Sholes design, including the offsets. Designers of laptops, with their fairly severe space limitations, have contrived to follow the Sholes offsets but to make interesting use of them by changing the shapes of the keys on the side edges of the keyboard, so as to fit into a rectangle. The laptop on which I am typing just now has a normal size shift key on the left, next to \ next to Z, but above that it has a one-and-a-half size Caps Lock key next to A, and a one-and-a-quarter Tab key next to Q. On the right, there is a Return key which is an upside down L shape covering two rows, and a one-and-three-quarter Backspace key. At the top is a row of smaller-than-standard function keys, so that more can be fitted in the row, with a few more at the bottom right.

Other languages

There is of course some variation between countries and languages. Languages that use the Roman alphabet do not have to do much to make Sholes work for them—maybe add a few accents or special characters. Non-Roman alphabets obviously need more drastic change; but it's really just a question of making substitutions. But what about non-alphabetic languages? Chinese, for example, has very many more characters than could possibly be represented on a Sholes-like keyboard.

There was a form of typewriter developed for Chinese. It consisted of a tray of several thousand embossed metal characters, each one in mirror image, like those on a traditional western typewriter. But in this case the characters are all separate, not attached to any part of the machinery. In order to type a character, the typist has to locate a movable frame above the correct character, then press a lever, which causes that character to be lifted out of the tray and struck against the ribbon and paper. Although it was possible to achieve quite fast typing speeds (if measured in words per minute), it required the typist to train for a couple of years.

Nowadays, in China as in the West, most such work is done on computers, with western (i.e. Sholes) keyboards. There are a couple of different ways of typing Chinese on a western keyboard (involving multiple keystrokes per character and/or menus), which obviously have to be learnt. But essentially this is very much easier than trying to construct a direct representation of Chinese on a keyboard.

What would Sholes think?

Let us return, for a moment, to the fantasy of communicating across time with Christopher Sholes.

If you were to go back to 1877 and explain to Sholes that his keyboard design will still be in use at the start of the next millennium, despite the fact that *every one* of the mechanical constraints that determined the design in the first place will have disappeared, he might be flattered but would probably think you a little crazy. If you were to add that a keyboard based on his design will be attached to practically every typewriter-like device in the world, *including China*—he would surely have no doubt that you were certifiably insane.

Yet, at the start of the third millennium CE, such is the case.

The concept of a character

Earlier, I referred to 'letters or other characters'. We have already seen the letters of the alphabet and the digits of the Arabic numbering systems as characters, and inventions such as the typewriter or the ASCII coding scheme give us extra characters such as punctuation marks and currency symbols. Every so often, an old symbol gets resurrected for an entirely new purpose. Consider for example the @ sign, which used to be used to indicate the unit price of some commodity ('2lbs margarine @ 4d per lb = 8d', lb being a pound weight and d being a penny in the old UK coinage). @ has now been taken over as the universal email address sign and for other uses.

One character for which ASCII has a code (though neither Morse nor Baudot did) is the famous inter-word space, which I discussed earlier in this chapter. In this respect ASCII, following Baudot before it and inspired by the typewriter, has somewhat extended the notion of a character. Following the space-bar on the typewriter (which is treated very much like an invisible letter), ASCII defines space as a 'printable' character, distinguished from 'control' characters like *newline* or *tab*. We have now become completely familiar with the idea that the space is just another character. Further, the ASCII distinction between printable and control characters now seems rather strange, at least to computer programmers. Even Tab or Newline is just another character, with its own key on the keyboard and its own code in the coding system.

As an aside, an immense source of confusion and problems with machines has resulted from the fact that despite ASCII, there has been no agreement on which character should be used to represent the end of a line. ASCII has two, defined as 'carriage return' (CR) and 'line feed' (LF), both terms again being relics of traditional typewriters—a CR moves back to the beginning of the line on the typed page, and an LF advances by one line down the page. Files on the Windows operating system have lines ending with CRLF, on the Unix system the convention is LF, and on the Apple Mac it used to be CR. And this is not an exhaustive list of the conventions that have been used!

ASCII also distinguishes clearly (as Morse did before it) between letters and numbers. The Sholes typewriter on the other hand had digit keys for 2-9 but not for zero or one; the typewriting convention was to use lower-case ell for one and upper-case oh for zero.

I have assumed, and the typewriter and the ASCII scheme both encourage me to assume, that there are well-defined, separable things called characters, not only in alphabetic systems but also in syllabaries and other writing schemes. This is something of an over-simplification; we can see oddities even within ASCII, and more so when we consider other languages than English. Actually, much the cleanest character system is the Chinese: each character is self-contained and occupies a square block on the page—beautifully simple, if you forget for a moment about the number of different characters.

One oddity in ASCII is that it has *two* codes for every English letter: lower and upper case. The ASCII code for 'A' is different from that for 'a'. There is some reason for this—although there are certain rules about when to use capitals and when to use lower case, these rules are not clear or unambiguous enough for us to leave the decision to a machine. So when we are typing, we use the shift key to indicate a capital letter, and the coding is done accordingly.

We could easily have decided to make the shift key a character in its own right, a control code saying to the machine 'Now go into upper case', either to apply only to the following character, or 'locking' and requiring a corresponding 'down-again' code (this latter method is used in the Baudot system to represent numbers). But we didn't. On the other hand we typically use exactly this method to represent font or typeface variations; I have to use some such convention in order to produce *italics* or **boldface** in this book. We don't have separate codes for Roman A, italic *A* and/or boldface **A**.

This decision has many ramifications. Think for example of how names are typically organised in a directory, or words in a dictionary or an index. Traditionally, we do not distinguish between upper and lower case when arranging things in dictionary order. Similarly we expect modern search engines not to distinguish. But these expectations require our machines to be told that (for some purposes at least) 'A' and 'a' are the same.

In Arabic, each letter has not two but four different forms. But here the rules are pretty clear: the four shapes occur when the letter is (1) at the beginning of a word, (2) at the end of a word, (3) elsewhere in a word, and (4) on its own. Although making a traditional typewriter do that would be hard, this decision can now safely be left to a machine, so only one code for the letter is necessary.

In English printing, we have some letters that are usually, in many type-faces, joined together—called ligatures. The most common examples are f l, represented as fl, and f i, represented as fi. The typesetting system I am using for this book will do these ligatures automatically for me (except that I have to tell it *not* to do so when I want to show the letters separately). In older books, you sometimes see other ligatures (for example s and t are sometimes joined), though most other ligatures have now died out. Also it is traditional to form a single character from an a followed by an e in some circumstances, for example 'archæology'. But this example is more tricky, for two reasons. First, it only applies to some words of Latin origin, it is not a general rule for when these letters occur together. Second, if it is ever encountered in modern English it is regarded as a ligature of the two letters a and e—dictionary order treats it as two separate letters. But in the Scandinavian languages and in Old English, this character is regarded as a letter in its own right, with a position in the alphabetical order distinct from the two component vowels.

As we explore other languages, we find many complex examples. In German there is a double-s symbol ß (nevertheless treated as two esses in alphabetical order). In Spanish we have a letter that is printed as ll (two ells), but is regarded as a single letter in its own right, with its own alphabetical position. Decorations on characters, such as accents or umlauts or cedillas, introduce their own complications. Sanskrit is written in an alphabetical system (the Devanagari script), but all the letters that make up a syllable are joined by ligature into a single syllable-symbol; there are hundreds of different ligatures. Sanskrit and Arabic also share the property that vowels are typically regarded as decorations on the consonants, rather than letters in their own right.

These complexities are hard to deal with in a coding system; eventually, instead of representing self-contained characters, some codes have to be used to represent instructions to the machine as to how to interpret the characters, or how to render them in readable form.

So the idea of a 'character' is a little complex. We English speakers and writers are lucky to be spared some of these complexities. And, just possibly, the development of computing in the English-speaking world benefited from the relative simplicity of our script.

6. Organising information

Every act of communication involves organising information—choosing what to communicate, and how to express it, whether in speech or writing or some other method. All forms of writing, even writing in order to enhance your own memory (for example, a shopping list), require organisation—of ideas, connections, facts, words, numbers, feelings, desires, intentions, stories, opinions, or whatever. We have already seen how the earliest forms of writing were for such purposes as commerce and administration, and such writing is necessarily an act of organisation. Another purpose, which developed early, probably counts as the first scientific endeavour: the study of the heavens.

Astronomy

Observation of the stars, more particularly systematic observation and recording, began very early in human history. Much of what we know about it derives from written sources from the first millennium BCE, particularly Babylonian clay tablets, but these certainly include material from much older sources, now lost. One particular set of observations of the planet Venus probably dates to the seventeenth century BCE

Such observational data might reasonably be termed 'information' precisely because it is systematically collected and organised for recording. In fact, it may now provide us with information not envisaged by its authors. Despite various uncertainties about the accuracy of the copies we have and the exact interpretations of the record, these observations can now be used to validate aspects of historical chronology, because our present astronomical knowledge allows us to determine the exact positions of the planets in the second millennium BCE.

Babylonian astronomers constructed extensive catalogues of stars and

    https://doi.org/10.11647/OBP.0225.06

constellations. We have copies of two such catalogues, the originals probably dating from around 1200 and 1000 BCE respectively.

Astronomical matters are of course important for human affairs. Sun,
moon and stars have been the most important resources for navigation
across open seas ever since humans tried such navigation—only in very recent history replaced by satellite navigation. Astronomical navigation, as
practised over the last two or three centuries, requires the preparation and
distribution of nautical almanacs containing tables indicating the positions
of sun, moon and 57 selected stars (as well as, famously, an accurate marine
chronometer or clock).

The *Computus*

For an earlier example of the perceived importance of astronomical data, one
of the questions that much exercised the early Christian church was when
to celebrate Easter. This question brought into existence an entire subject of
study called the *Computus*, concerned with the various astronomical events
and cycles by which calendars are determined. Proper calculation of the
date of Easter requires the taking into account of the length of the true solar year (approximately 365-and-a-quarter days—but the quarter is not exact), the true lunar month (again approximately 29-and-a-half days), and
the week of seven days. The length of the solar year (then assumed to be
365-and-a-quarter days exactly) had been the basis for the introduction of
the Julian calendar under Julius Caesar in the first century BCE. Various
different versions of the Easter calculation were defined, but the one that
came to dominate was formalized by the Venerable Bede in the eighth century, following a formula devised by Dionysius Exiguus in the sixth. Bede's
great work on the Computus, *On the Reckoning of Time*, contains a number of
tables based on astronomical predictions, and shows the date of Easter for
many years in the future.

Much later, in the sixteenth century, the Gregorian calendar was introduced by Pope Gregory. The difference between the Julian and Gregorian calendars is to do with the difference between the assumed 365-and-
a-quarter days and the true length of the solar year. But the specific reason
for its introduction was to readjust the date of Easter in relation to the sea-
sons, in particular to the spring equinox, to what it had been at the begin-

ning of the Christian era. Currently, the date of Easter as celebrated in most western churches differs from that used in most Orthodox churches. This is a consequence of the fact that the western churches generally converted to the Gregorian calendar, while the Orthodox churches stuck to the Julian calendar.

Tax collection

Another early example of information organisation was to do with taxation.

We know that there was a system of taxation in Egypt, early in the Old Kingdom, about 3000–2800 BCE. The easiest people to tax are the farmers, because typically both their means of production (fields and livestock) and what they produce are clearly visible to all. So the principle might be that 10% of the crop goes to the local governor or tax collector. Except that this is hard to police—you would have to have someone watching the farmer all the time. But you can measure his fields once or at long intervals, and count his livestock also. For the fields, you might assume that a field of a certain size will have a certain yield in a year, and tax the farmer on that basis. Maybe you need to distinguish between the very productive fields located in the Nile flood plain, and the somewhat less fertile fields on the hills. Then the farmer can be taxed not on what he actually produces, but on what the system assumes that he produces.

All of which requires the tax collector to keep records, in a standardised form. What area of fields, in each yield category, does this named farmer have? At once we see not only that the messy world has been manipulated into a tidy form, but also that this manipulation is not neutral. It is to the advantage of the farmer that his field on the edge of the slopes is classified as 'hill'—but to the tax collector, the advantage is reversed. Since the tax collector is the literate one who actually makes and keeps the records, his view is likely to prevail!

Census

One of the things a tax-collector needs to know is who the tax-payers are, and what they own. Governments have been conducting censuses as long as they have been systematically collecting taxes. There are of course other purposes for conducting a census—knowing who to call for military service, all sorts of planning exercises that need statistical data, and so on. The word itself is Latin, and in Rome originally signified a list of those available for military service. But the concept is probably at least as old as tax collecting.

In England in the eleventh century, for example, William the Conqueror initiated a census of all his possessions, people included, called the Domesday (Doomsday) book. The book is primarily organised around land—the rural estates. In such feudal times, the people come with the land. But it includes the names (first names only) of under-tenants of the lord of the manor.

Modern censuses are normally tied to notions of 'residence' and 'household'. A return is made for each household, and includes every person resident in that household. Both these notions are fuzzy at the edges. Nevertheless, the requirements of census-taking have played an important role in the development of ideas of information processing, as we shall see in Chapter 11.

History

In early human societies, history and mythology are irretrievably intertwined. One might argue that the same is true today, as in the saying attributed to Winston Churchill, that 'history is written by the victors'. Nevertheless, we now associate the great classical Greek historians of the fifth century BCE, Herodotus and Thucydides, with the attempt to put history onto a more systematic footing, and to base it on carefully gathered evidence, in the process distinguishing history from mythology. Although I started this book by arguing that recorded history could not begin until we had developed writing, it is clear that this is not sufficient—we don't immediately start the systematic recording of history because we have invented writing. These two Greeks had significant predecessors concerning whom less is known; but their role in developing historiography, the systematic study of history, is clear. Although they differed as to emphasis, between them they championed the meticulous gathering, analysis and evaluation of evidence, from witnesses and documents, about the events and circumstances they wanted to describe.

Libraries

We have already seen in Chapter 4 the importance of libraries in our story, as a method of communication. They also play a central role in methods of organisation of information.

Consider for example the great classical libraries that I mentioned: the Library at Alexandria, for example, or the House of Wisdom in Baghdad, or the library of one of the big medieval monasteries. In all these cases, scholars would arrive from remote places hoping to find enlightenment of some kind. The Alexandria library, for example, might have contained hundreds of thousands of items (the collection seems to have consisted mainly or entirely of papyrus scrolls; a single work might take up multiple scrolls). Either locating particular known items, or looking for multiple items on a subject, would have been a far from trivial task. The library was arranged by subject, each subject having a bin to contain the collection of scrolls. A tablet above the bin listed the contents of the bin, and each scroll had a tag attached to it, giving the author and subject. This kind of information was also the basis for what is supposed to be the first library catalogue, produced by a librarian called Callimachus for some of the material in the Alexandria library in the third century BCE. Just to indicate the scale of the finding problem, the catalogue ran to 120 scrolls.

The art of the library catalogue (thinking now of the present) is of interest to us for two reasons. The first is that it provides an organisation of the books or other materials in the library. It does this by collecting information or data *about* each book (sometimes referred to as metadata), specifying for example its author or authors, its title, when and where it was published, some codification of its subject matter, etc. It then provides access tools so that a book can be identified in a variety of ways, say by looking up the author. The location of a book on a shelf, quite likely as part of a subject arrangement, provides one (but only one) way of finding it. A catalogue typically provides multiple ways, suitable for different forms or types of enquiry. How it does this depends on other technologies available. Before the availability of computer-based library catalogues, various forms of index were needed—some were on cards, some printed on paper.

The second reason we may be interested in library catalogues is because of the organisation of the catalogue data itself. Consider for example the data elements suggested above (author, title, publisher, date, subject). If an index based on any of these elements is required, each has to be treated in a consistent fashion across different items. For example, in order to make it easy (or even possible) to look up an author name in an index, the recording of the author name must follow a well-defined format and set of rules—and (ideally) be consistent if the same author has written multiple books. This might well require some manipulation of the messy real world.

Just for example, I have a book on the shelf next to me by the great physicist and Nobel prize-winner, Richard Feynman. Well, actually, the author's name appears as Richard P. Feynman. Elsewhere (not in this book) it is possible to discover that his middle name is Phillips. Another book, containing a collection of his writings, is titled *No Ordinary Genius: The Illustrated Richard Feynman*—which gives as author Richard Phillips Feynman (together with another person as editor). All of this would probably not matter very much, since author indexes are normally ordered by surname, and I would be quite likely to find entries for *Feynman, Richard*; *Feynman, Richard P.*; and *Feynman, Richard Phillips* quite close to each other. Besides, Feynman is a relatively uncommon name. And as I have only one Feynman in this book, I can get away with *Feynman, P.* in the Person index at the back. But names can cause much more serious problems than this—some further discussion below.

Forms

A very particular kind of organisation is required when you have to complete a form, whether on paper or online. Every time you fill in a form, you are slotting information that you have, about yourself and the world around you, into a kind of information-organisation devised by someone else. However messy the world around you, or the information that you have about it, the form makes you think about it in a particular way.

Let's take a name, for example. You have a name. More than that, I can say with some confidence (and any form you have to complete may well assume) that you have a surname that you inherited from your parents, or perhaps acquired later by marriage, and one or more given names—but if more than one, it's probably only the first that you actually use. So the slot in the form into which you are supposed to put your single used given name is quite likely labelled 'First name' (in my childhood, it was often labelled

'Christian name', though that obviously culturally biased terminology has largely disappeared). A form in the USA might ask for a 'Middle initial'.

But the entirety of the structure is culturally biased, of course. Chinese people coming to the West typically learn to reverse their two names—because by default in China, the surname comes first. Someone from the Indian subcontinent (I have friends like this) may have acquired only a single name as a child, and have had to invent a second for the purpose of filling in forms and (more generally) living in the west. I have several relatives who have two given names but actually use the second. I also have friends and relatives with double-barrelled surnames, not hyphenated but spaced, like the composer Ralph Vaughan Williams—that's not a problem when they complete a form themselves, but is definitely a problem for the library cataloguer. Other parts of the world have different practices—for example, in both Spain and Portugal, most people have double surnames. And of course if we go back in history as well as elsewhere in geography, the range of variations is huge. Often, the messy world has to be doctored in order to fit into a tidy form. And this is only the very first bit of the form!

Addresses

The next question on your form, after your name, is quite likely to be your address—though like your name, the form may require it to be split into multiple parts. This in itself is a slightly strange requirement in this day and age. If a friend writes down an address for me on a piece of paper, I will probably have no difficulty in parsing it—in distinguishing the house number, the street name, the town name and the postcode. It's a simple enough process, bound by rules, considerably simpler than long division—so I would expect a machine to be able to do it reasonably well. (If the form you are completing is not online already, it's likely to be fed into a machine shortly after completion.) So why isn't it left to the machine to do the parsing?

One reason might be that the form of an address is quite strongly history- and culture-dependent. More specifically, the national postal systems discussed in Chapter 2 have been very closely involved in the determination of standard address forms. Thus the standard varies considerably from country to country. Furthermore, although there have been attempts (since the establishment of the Universal Postal Union in 1874) to define an international standard format for postal addresses, these now seem to have been abandoned. From the point of view of post alone, it probably doesn't matter very much—the Universal Postal Union is a federal structure, so as long as the postal service where the letter is posted can recognise the destination country, it can leave the rest of the address for interpretation by the local postal service in that country. However, it can cause problems for other uses of addresses (of which there are many).

One prime current example is the postcode. Although the first divisions of large cities into postal regions began in the nineteenth century (London 1857), and some more detailed attempts began in the 1930s, these mostly originate from the 1960s and '70s, a period that might just still be regarded as the heyday of the post, but perhaps its tail end. Many postcode systems provide a rather coarse level of granularity, a district containing many houses, but some are much more precise. In the UK system, for example, a postcode does not uniquely identify an address, but specifies a small group, up to a hundred but probably many fewer.

Postcodes (indeed, addresses generally) serve or contribute to a number of different functions other than postal deliveries. For example, postcodes are commonly used for satellite navigation (despite the fact that they were mostly devised before satellite navigation was invented). But for this purpose, one would like a very fine granularity. On the whole, the UK system works very well for this purpose, particularly in cities, but sometimes in the countryside it is not precise enough. But there is considerable variation between countries, even among those countries that have postcodes.

Returning to the parsing question raised above: what we do now find commonly in the UK is that the form-filler is invited to provide *only* the postcode, and then allow the computer to deduce almost fully the rest of the address, offering the user a small choice of house numbers and possibly street names. This deduction is based on a database, to which the computer has access, of postcodes and corresponding full addresses. To make sense of this statement, we need to talk a little about databases.

The concept of a database

Despite my comments above about the difficulties inherent in both, name and address data is often held up as a good example of a kind of information with a high degree of structure, a high degree of regularity, and a high degree of consistency. As a result, it is taken to be a good candidate for storage in a computer in what is commonly known as a database. If you keep your contacts on your computer or your phone or both, they will be held in a database. This means that even if some of the addresses take a slightly different form from others, or some data is missing from some, they are all held in a common structure. There are several reasons for doing it this way: essentially they revolve around how such data can be processed automatically, including for display to you. Thus, for example, you would expect to be able to see an alphabetical list of names. Once again, alphabetical sorting of names is not quite as straightforward as it might seem; nevertheless, you probably expect your computer, and your phone if it is even remotely clever, to be able to do that.

Databases, and computer programs that manipulate databases, are staples in the world of computing. Indeed, the maintenance and manipulation of databases is a vastly more important function of computers than calculation. Consider, for example, the computers in your bank, which look after your bank account. Clearly they have to do some calculation, when you add or withdraw funds or move them around—but by far their most important function is to maintain consistent records of all such transactions, as well as all the other information relating to this and every other account. Furthermore, you have probably never seen those computers make arithmetical mistakes—that's the easy part—but you are quite likely to have seen instances where, for one reason or another, transactions have gone AWOL.

If you frequently do online transfers, and have ever made a mistake, you may have discovered that a mistake in the destination account number can be much worse than a mistake in the amount. An account number is not really a number at all (nobody ever needs to do arithmetic with account numbers): it's a code identifying a particular set of database entries. As you may have read in many newspaper reports, if you transfer money into the wrong account, and the account holder is not willing to do anything about it, neither you nor your bank can recover the money. Until very recently, in the UK at least, banks in these circumstances did not typically check names, only account numbers. This is probably because of all the issues discussed above with respect to names. If you do not know the exact form of the name of your payee, as held in the bank's database, then the chances are high that you would enter it in a slightly different form, so the banks prefer to rely on the code. Nevertheless, it is much easier for a human being to make a mistake with a long numerical code than with a name, so this logic can be counter-productive.

Varieties of database

Databases come in many different forms. In the present day, a database is generally assumed to be held on a computer. Many such systems follow the principle that data should be divided into its smallest coherent component parts, and that exact rules of inference should be specified, completely determining what can be learnt by recombining the data elements in new ways. This is a reductionist view, and has a strong analogy to the status of arithmetical calculation ever since the rules for this were codified. Some kinds of data are amenable to this approach, and it brings advantages in the ability to manipulate it in well-understood ways. However, not all data, let alone all information, can be treated in this way.

In the past, long before computers or the coinage of the word *database*, we have seen many collections of information that would now be called databases. Of those we have discussed in this chapter, all collections of completed forms, all library catalogues, all sets of census returns (and all tables derived from them), all tax collectors' records of the people and institutions that they tax, all banks' records of people, accounts, transactions, and so on and so on, can be seen as databases. All involve rules of organisation and of manipulation.

We return to the theme of calculation in Chapter 10, and the broader theme of information processing in Chapter 11.

7. Picture and sound

For this chapter and the following two, we leave aside for the moment the world of writing and characters, alphabets and numerical digits, to consider pictures and sound.

As we saw in Chapter 1, pictures played an early role in the development of writing systems. However, at that point the paths split apart. When it comes to recording information in a form other than conventional language, such as diagrams or images or pictures, moving images, and sound, we have (broadly speaking) bypassed the language-writing-alphabet-digital code sequence. But pictures, in particular, have their own sequence that leads from the development of photography in the first half of the nineteenth century, to the almost universal digitisation of everything today.

Most people today are aware that pictures are usually made up of dots (*pixels*), in a rectangular array—cameras are typically sold on the number of pixels they have. If you look through a magnifying glass at a picture printed in a magazine, you can see the pixels. Below, I will describe the sequence of development from the first photographs to today's pixel-based images. However, first it is worth exploring a couple of byways of image-making.

Art

A very early art form, visible for example in the cities of Pompeii and Herculaneum, which were buried in volcanic ash in the first century CE, uses the method of mosaic. In mosaic, a picture is built up out of small tiles; each tile is a single colour. The subtlety that can be expressed by such means is astonishing. However, there is a significant difference between the layout of a mosaic and the modern notion of digitised images. In the modern version, the picture is built on a rectangular grid—the colour at each intersection of the grid is recorded. In mosaic, typically the tiles (even if they are often

 https://doi.org/10.11647/OBP.0225.07

square individually) are usually laid following the shape of the design, so
that a curved boundary in the picture is usually made of two curved lines
of tiles (see for example Figure 3).

Figure 3: Ancient Roman mosaic, featuring a panther—detail
from Pompeii, National Archaeological Museum, Naples
`https://commons.wikimedia.org/wiki/File:`
`MANNapoli_SN_mosaic_Panthere.jpg`
Public domain.

Very much later, in the late nineteenth century, a small group of artists
(Georges Seurat and others) developed a technique of painting relying on
small dots of colour. Rather than creating a particular desired colour by
mixing a small number of basic or primary colours on a palette, they would
use the same basic colours individually, close to each other in small dots, so
that to the eye they would appear to mix into the desired colour. The method
was known as *pointillism*. One of its effects is supposed to be to brighten the
colour sense of the viewer (see Figure 4).

Weaving

A simple form of weaving involves a set of parallel warp threads stretched
over a frame, a mechanism to lift alternate threads away from the others,
and a shuttle to pass the weft thread from one side to the other, in between
the lifted warp threads and those left flat. Before the next pass, the lifted
threads are returned to level and the remaining threads lifted instead.

Figure 4: Georges Seurat's *Honfleur, un soir, embouchure de la
Seine* (detail) Museum of Modern Art, New York
https://commons.wikimedia.org/wiki/File:Georges-Pierre_Seurat_-_
Honfleur,_un_soir,_embouchure_de_la_Seine_-_Google_Art_Project.jpg
Public domain.

If you want to weave a coloured pattern of any kind, the warp and/or
weft threads can be of different colours. But to weave a complex pattern,
even a picture, requires more complex control. One way to do this is as follows. The weft thread is a single neutral colour, but the warp threads are set
as a sequence of colours, as it might be yellow / green / red / blue, repeated
over the width of the fabric. Then the colour of a particular location in the
woven fabric, on the front or top face, is determined by which warp threads
have been lifted (and are visible) and which have been left flat. As in pointil-
list painting, if the threads are close enough together, the eye perceives the
mixture of the selected basic colours as a new colour.

A method like this gave rise to a fascinating invention at the start of the
nineteenth century, due to Joseph-Marie Jacquard. He devised a system for
controlling a loom, using a sequence of punched pasteboard cards, one for
each pass of the shuttle. The positions of the punched holes in each card
controlled which of the warp threads were to be lifted for the next pass. A
sequence of cards could be punched, potentially of any degree of complexity,
to produce a complex design in the fabric. The cards with punched holes can
be seen now as a digitised representation of the finished design (see Figure

5).

Each punched card determines the pattern of lifted warp threads for a single pass of the shuttle.

Figure 5: A modern Jacquard loom
https://commons.wikimedia.org/wiki/File:India_-_Sights_%26_Culture_-_
Hand-loom_pattern_cards_for_silk_sari_weaving_(2507306023).jpg
CC BY 2.0.

Jacquard's loom was a major development in weaving technology. Predating photography, it continued in use for almost two centuries, in parallel with but independently of the developments in photographic imaging described below. Today's computer-controlled looms are direct descendants of Jacquard's invention, but finally now integrated into the rest of the digital world. However, Jacquard's use of punched cards was to inspire two quite different developments, as we shall see in later chapters. James Essinger, in *Jacquard's Web*, argues convincingly for Jacquard's seminal position in the development of information technologies.

Photography

The idea of photography, using light-sensitive chemicals, was developed in the first half of the nineteenth century. The best-known name from that time, Louis Daguerre, invented a form of photograph called the *daguerreotype* in

1839, but a much more successful and long-lived method was the *calotype* of William Henry Fox Talbot (1841). This was the forerunner of the negative process that was the norm for photography right up until the development of the digital camera at the end of the twentieth century.

As an aside, Talbot's name is a great example of the kind of issue with names that I discussed in the previous chapter. He is usually referred to as 'Fox Talbot', on the assumption that this was a double-barrelled but unhyphenated surname—he was referred to thus even in his lifetime. Actually, he did not regard Fox, a family name of his mother's, as part of his surname. In addition, he normally used the 'first' name Henry, not William.

Photography in this form can be described an *analogue* process. We make a distinction between *analogue* and *digital*: in an analogue representation of something, a *continuous* or smooth variable in the outside world is represented by a continuous or smooth variable in the system. In the case of chemically-based photography, light causes a chemical change on the plate—more light causes more change, in a more-or-less smoothly varying way.

At this point, someone familiar with traditional chemical photography might well object 'What about grain and graininess?'. It's true that at a fine level, the chemical process is not smooth. But this is not a matter of design: it's rather an accident of the process. It's when we start designing in granularity that we call something 'digital'.

When we deliberately slice up the world, or some smoothly varying part of the world, into discrete parts, in order to measure or represent or process it, we are beginning the process of digitisation. As we will see below, some processes are mixed, in that they include some continous elements that have been divided into discrete parts and some that have not—this is particularly so in the development from simple monochrome chemical photography to today's digital images.

However, in order to make this process a little clearer, we start with a simpler example: sound.

Sound

As with so many things, digitised sound predates the computer era, though not by very much. The standard form of digitised sound, *pulse code modula-*

tion, was invented by Alec Reeves in 1937 and is still in use today.

Sound is waves of pressure moving through the air or other medium. A pure musical note is a simple sinusoidal wave of fixed frequency (middle C is about 260 Hertz, or cycles per second); all other sounds are more complex patterns of pressure change. One can draw a graph of how pressure varies over time—it looks like an undulating line (see Figure 6). A microphone (invented in the nineteenth century) turns the pressure waves into electrical waves, in an analogue fashion, so that the pattern of varying electrical current looks like the pattern of varying pressure in the air.

In order to digitise a continuous curve like this, we need to make two separate quantities, which are naturally continuous, into discrete lumps. The two quantities are time and pressure. So first we divide up time: we choose discrete time intervals—in the case of CD audio, 44,000 times a second. Then we measure and discretise the pressure at each of those times. Again, in CD audio, the discrete pressure levels are defined with 16 bits—that is, there are 2^{16} (which is about 65,000) pressure ranges. The result is a representation that can be thought of as turning the smooth pressure curve into a stepped line, with very tiny steps—again, the diagram shows a rough picture. On playback, the player has to recreate (to a close approximation) the smooth curve from this digitised step curve.

This is pulse-code modulation, and is all we need to digitise a single sound channel. For stereo sound (again as on CDs), we need two separate channels like this (more on stereo sound in the next chapter). The invention of the CD as a digital recording medium in the late 1970s allowed the recorded music industry to make a relatively smooth transition into the digital era—though the relationship between music more generally and the digital world is more complex than this, as discussed further in Chapter 9.

Now let us turn back to photography and its developments.

Moving images

A photograph shows a static image, a snapshot in time. How do we turn this into a moving image?

Probably you already know the answer to this question—and in any case, it is very like part of what I described above for sound. Although time is, in principle, a continuous variable, we can treat it as discrete by taking very

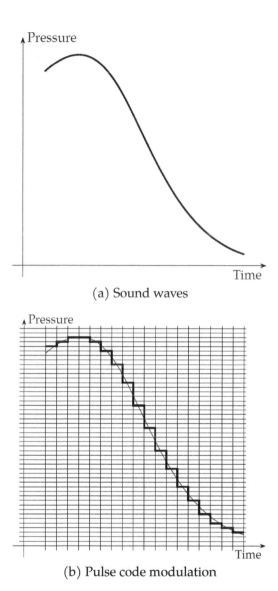

(a) Sound waves

(b) Pulse code modulation

Figure 6: Sound. Diagram: the author.

small steps in time, and at each of these time steps, we take a photograph. Actually the steps do not need to be nearly as small as for sound—film typically works at just 24 frames a second. If you see an image changing in small steps 24 times a second, it looks pretty much as if it is moving smoothly. Mostly, you probably aren't aware of the step-changes.

This principle, on which film is based, was understood in Victorian times—in museums you will sometimes see Victorian toys using the same idea, for example a spinning drum with slits. If you look at it from the side, each slit gives you a momentary view of a drawing; the next, which appears a moment later, is like its predecessor but with small changes. The impression is of movement. Actually making a film camera was somewhat tricky, but in the 1870s Eadweard Muybridge demonstrated the idea with a series of photographs of a galloping horse. The photographs were taken by separate still cameras, on the side of the course, but the effect is of a single film camera tracking the horse. Muybridge developed a device to show the moving image, the Zoopraxiscope, very like the toy described above.

The first patent for a cine camera was due to Louis Le Prince in 1888, and the process became practical in the 1890s. Notice that it is only the time variable that we are treating discretely. Ordinary old-fashioned celluloid film retains the ordinary analogue process for each individual frame.

Colour

In parallel with the challenge of recording moving images, Victorian inventors concerned themselves with that of developing colour photography.

The basic principle is based on the idea of primary colours. This idea had been around from at least the seventeenth century, and a specific theory concerning human perception was put forward by Thomas Young in 1802, further developed by Hermann von Helmholtz in 1850, and proposed for photography by the physicist James Clerk Maxwell in 1855. The idea was that any colours can be made up from a small number, probably three, of primary colours. The artists' primaries are normally taken to be red, yellow and blue. Actually the idea of primary colours is somewhat complex and messy; I will return to it below. But if we assume in the meantime that the idea is good, then we have in principle a natural way to build a colour photograph—take three separate photographs, with red, yellow and blue filters, and then print the three resulting images, in red yellow and blue, superimposed. In terms of the ideas discussed above, the full continuous colour spectrum can be reduced to just three discrete components.

But once again, actually doing this is a little tricky. Maxwell's attempt in 1861 was not very successful: the idea had to wait until the 1890s to be developed into an even remotely practical form, and till 1907 for a truly commercial process. In the Autochrome system introduced by the Lumière brothers in 1907, the photographic plate included an integral screen of small dyed potato starch grains, distributed irregularly, but small enough that the eye would not distinguish individual grains. The area of plate under a coloured grain would respond to that coloured light, and not to the colours that had been filtered out. When developed (including a reversal process to get back from a negative to a positive image), each area of the image would be seen through the correct coloured filter, the same that it had been exposed through. Despite a number of disadvantages, this was a successful system that lasted until the 1950s.

In the 1930s, two successful classical musicians working for Kodak developed a process (Kodachrome) with different light-sensitive emulsions, responding to red, green and blue light respectively, all on the same plate. This became essentially the dominant system of colour photography (both still photography and cine film) until digital took over.

Other Victorian ideas

Another idea that emerged in Victorian times was that of 3D images. Here the basic principle predates photography: Charles Wheatstone demonstrated a stereoscope in 1838. It was a device that presented each eye with a picture (he used drawings), the two pictures being slightly different, in a way which caused the viewer to merge the two into a three-dimensional scene. Although some of our depth perception comes in other ways, an important component comes from our binocular vision, and Wheatstone's device both relies on and demonstrates this fact.

Almost as soon as it was possible to do so, various systems using two cameras were developed, allowing the viewer to see commercially-prepared still 3D photographic images of places and scenes. There are several different methods of causing the images presented to the two eyes to differ, but the principle is the same, right up to and including modern 3D film and television.

Another concern was to be able to print photographs in newspapers. Newspaper ink is either on or off—it is not possible to print shades of grey directly in the newspaper printing process. Newspapers of the time sometimes used etchings, which could be prepared with a lot of effort, and could represent greys by hatching and other devices; but to produce an etching from a photograph (essentially an artist's copy of the photograph) was not ideal. Already in the 1850s, Talbot had an idea for how to produce a printing plate directly from a photograph, although once again it took a little while to be made practical. The idea is to reduce the picture to a grid of black dots, varying in size. In areas of the picture where the dots are small, the eye sees mostly the white paper; in those areas where the dots are large, the eye sees dark grey. The first printed halftone pictures appeared in the 1870s, and in the 80s commercially successful methods were developed. Similar methods are still in use, including for printing colour pictures.

Facsimile transmission

We have seen how. for some purposes to do with images, we need to *discretise* some smoothly varying part of the image. We have already discretised both time (for cine film) and the colour spectrum (for colour images); now we need to turn our attention to other smoothnesses.

A (two-dimensional) picture or image has... well, two dimensions. If you are at all familiar with graphs, then you might think of them as the x and y dimensions. x is the left-right component, and y is the up-down component. Thinking for the moment only of monochrome still images, over each of these dimensions, the picture may vary anywhere between black and white, that is it may take any shade of grey. This (brightness) represents a third smooth variable to add to the two dimensions.

If you want to transmit anything over a telephone line or a radio channel, you have essentially only two smooth variables to play with: the level of the signal, and time. For (single channel) sound, this is all you need: as we saw above, a microphone turns time and air pressure variation into time and electrical current variation, which can be transmitted over a telephone wire or over a radio channel. But for pictures, we have a third variable. It follows that we must do something different with one of them.

Once again, the Victorians identified the challenge—the first serious experiments in fax transmission were by a Scotsman, Alexander Bain, in the 1840s (that is, well before the invention of the telephone!). His basic idea, which remained the dominant method right up to modern fax machines, was to *scan* the original, in a line from side to side, and then move down a small distance and scan a new line very close to the previous one. The whole image would be covered by a succession of lines.

Bain's originals had to be specially prepared for this purpose, and would only deal in black-and-white. It would take quite a few more years, to the 1880s, before one could scan an existing photograph, in shades of grey, and longer still for any kind of commercial fax transmission. Nevertheless, there were successful commercial systems in the 1920s.

We now see that what Bain did was to discretise the y dimension, leaving the x dimension to be represented in analogue by time, and the brightness in analogue by the level of the signal.

Television

So the next thing to invent is the transmission (along wires or over radio) of moving pictures. But we already have the ingredients for this: if we discretise both the y dimension and the time dimension of the moving image, we are left with just two dimensions needing analogue representation.

The process of line-by-line scanning for moving images has a name—*raster* scan. (For some reason this word is not usually used in the fax context, though the principle is exactly the same.) In an old-fashioned television tube, the cathode ray beam that causes the phosphor on the screen to glow follows the same raster process to regenerate the image.

The idea of using such a method for television transmission is also Victorian. An electromechanical method of raster scanning was patented (but not actually developed) in the 1880s. Again, development of commercial systems (as opposed to demonstrations) took a little longer; the name most closely associated with the invention of television is John Logie Baird, in the 1920s and 30s. The BBC made its first television transmission in 1929, and began regular transmissions in 1932. Initially the raster scan was based on electromechanical processes, and used 30 lines per image, building up to 240 lines in 1936. But electronic methods were already in the air, and in the same year the BBC switched to a 405-line electronic system. This was replaced in the 1960s by a 625-line version.

How about colour television? Again, we already know in principle how to do it—we need three colour components, so each frame has to be transmitted three times, once for each colour. Various schemes were devised to do this, but commercial systems started in the 1950s.

There is an interesting problem at the display end. We now have a cathode ray tube with three separate cathode ray guns constructing a raster pattern on the screen, which now has to have three different coloured phosphors. How to ensure that the right ray hits only the right phosphor? The usual way to do this is to have the phosphor in closely packed dots, and a metal filter screen just behind it, with a pattern of holes. These are aligned so that the ray corresponding to red can only reach the display screen where there are red phosphor dots, and so on. We can think of this as a sort of last-minute discretisation of the x dimension.

Full digitisation

Now, of course, almost all of our sound (recorded and/or transmitted), and almost all of our still and moving images (ditto), are fully digitised. Digital cameras, both still and moving, contain grids of tiny photo-receptors, which record light intensity digitally in three colours. Display and printing devices turn pixel-based data back into visible images. In between, all sorts of processing can take place, entirely in the digital domain. Telephones, radio, and recorded sound are similarly treated. Your mobile phone digitises the sound itself, and received digital sound from the other end. Your landline probably doesn't yet—it sends an analogue signal to the exchange, where it is probably digitised, and receives an analogue signal back.

If you go back in your time machine to the 1840s, to fetch Talbot and bring him forward in time, stop first in the 1980s. At this point he would be totally astonished by many things, including miniaturisation and mechanisation, although if you showed him the machine at your local pharmacy which was used to develop and print your photos, he would be able to get some grasp of the chemical and optical processes involved. But if you bring him here to the third decade of the twenty-first century, he would recognise nothing whatever of any part of the process, at any stage between the camera lens and the display on your screen.

8. On physics and physiology

In Isaac Asimov's *Foundation* trilogy, written in the 1940s but about the distant future, he describes a device that plays a recording of a 3D moving image with sound. The display device is a glass cube, in which the viewer sees a human figure talking—something like a talking head on television, but in 3D. Asimov does not explicitly say this, but the impression is that the image inhabits the 3D space inside the cube; people can watch from all around, but the people at the back will see only the back of the figure. Asimov was a biochemist, but one might describe this as a physicist's version of 3D film. This is in complete contrast to Wheatstone's original 1838 stereoscope, and to modern 3D film and television, and indeed to virtually everything tried in between these two dates, which might be described as relying on physiology—on the fact that we perceive depth through our binocular vision.

In this chapter I want to explore this space a little. I will return to 3D vision later, as in the previous chapter, I will start with sound.

3D sound

Our sense of the location of the source of a sound depends in part on the fact that we have two ears. The difference between what our ears hear and report to the brain allows us some degree of directional sense of where the sound is coming from. This is the basis for stereo sound systems. Given different sounds from two separated loudspeakers in a room, our ears can have some illusion of sound location.

However, this illusion is not very good. The sounds delivered to our two ears by two loudspeakers (in a room with its own aural character) are only a very rough approximation to what might be heard in a real environment with real sound sources, and of course real echoes from whatever else is in that environment. So what might be a better way?

   https://doi.org/10.11647/OBP.0225.08

There are two ways to go. One of them is to have many more loud-speakers, potentially with different signals to each one. 'Surround sound' systems, used for example in cinemas, are a move in that direction. But it could go further. I once came upon a public performance of a recording of a 40-part motet. In a large empty hall at the back of a church, there were 40 loudspeakers, each mounted on a stand at head height, distributed in a rough circle around a room. I could wander around and in and out while the music was playing, hearing in different ways, for example concentrating on one or a small group of parts, with the rest in the background. Exactly what I heard at any point depended on which way I was facing as well as my location. In addition to the different relative location of the ears as one turns, our ears are themselves each to some extent directional, and one's head casts an aural shadow.

That's a true physical attempt at a solution to the problem. However, it's not a feasible general approach to hi-fi in the living room!

The other direction would be physiological. We can take much more seriously the idea of delivering different signals to each ear—in fact good headphones make for a much cleaner aural environment, each ear hearing only its own signal, with no interference or cross-over and no echoes. However, in order to do this properly, the recording should be made in a similar fashion. That is, one should use a pair of microphones, each in its own shell-like mount, on either side of a head-shaped object.

This is known as *binaural* or *dummy head* recording, and is quite different from normal stereo recording. It is seriously difficult to do well. For one thing, everyone's head is a different shape, as are their ears and ear canals. For another, if the listener moves or turns their head while listening, the dummy head that was doing the recording did not move in the same way at the corresponding time during recording, so at this point the listener's experience will be distorted. Binaural recording cannot be a general solution, any more than multiple speakers can be.

Thus ordinary stereo and surround sound occupy a slightly uneasy place somewhere in between a true physical solution and a true physiological one. This is not to say that sound recording and playback is necessarily bad—some things come across wonderfully. But it is, necessarily to some degree, a distortion of the original sound.

The physics of colour

If you pass a bright white light through a prism, onto a white surface, you get a display of the spectrum of colours, as in a rainbow. This phenomenon was studied by Isaac Newton in the seventeenth century, but was not fully understood until the the nineteenth. The visible light spectrum is now understood to be a part of a much larger spectrum encompassing all electromagnetic waves, including radio, microwave, x-rays, and gamma rays, as well as those just outside the visible range, called infrared and ultraviolet.

Light and the rest are wave forms, which can be characterised by their wavelengths; the spectrum shows all the different wavelengths. Visible light has wavelengths between approximately 380 nanometres (violet) to 750nm (red). White light normally contains a full range of colours. A surface may reflect light of different wavelengths to different degrees—then the surface may be perceived as coloured. Usually this would be a smear across some range of the spectrum. Also a light source may generate different mixtures of colours. Old-fashioned filament lightbulbs typically produce light that is stronger at the red end of the spectrum than daylight. Modern bulbs can often be made to emulate the old-fashioned ones or to be close to daylight—these are currently referred to as *warm white* and *daylight* respectively, with *cool white* somewhere in between.

One case that will be useful for further discussion is the sodium lamp, often used for streetlights. It's unusual in that the light it produces is (to a close approximation) strictly monochromatic—that is, of only a single wavelength, around 590nm, pretty well at the yellow-orange boundary. If the only light source in a scene is a sodium lamp, it is impossible to distinguish the colour of any surface, because however much or little light is reflected from a surface, all of it is this single colour.

The physiology of colour perception

The light-sensitive cells in our eyes are of two types, rods and cones. Rods do not distinguish colours; however, cones are further subdivided into three types with different colour sensitivities, which enable us to see colour. These are called *red*, *green* and *blue* cones, which is an approximate way to describe their respective sensitivity to different colours of the spectrum. But actually, each type responds to a smear of different wavelengths, and these smears overlap considerably.

If our eyes are presented with monochromatic light (such as from a sodium lamp), the response of each type depends on whereabouts in its smeared response the monochromatic light lies. Sodium light lies quite close to the peak of the response-smear of the red cells, but with a significant green-cell response as well (very little blue-cell response). Our colour perception depends on the ratios or proportions of these different responses—the brain says '*this* much red, together with *this* much green, but very little blue, looks like a particularly virulent yellow-orange'. It is *only* through these proportions that we perceive colour.

If the light hitting our eyes were not monochromatic, but smeared over a range of wavelengths towards the red end, we might nevertheless get a very similar effect: that is, a very similar proportional response from the three different types of cones. There are actually very many different combinations of the basic wavelengths that our eyes are quite incapable of distinguishing.

Three-colour theory

Given that our eyes only have the three types of cones to distinguish colours, it seems plausible that we can construct colours using three primaries. That is, it should be possible to fool the eye into thinking it is seeing any particular colour by presenting it with suitable combinations of the three primaries.

What do we need for primary colours? The colour cones suggest something like red, green and blue. Indeed this is what is normally used for what is called *additive* colours. If you start with red, green and blue light sources, you can generate white light and more or less a full range of colours. Exactly this is done in some projection systems, with three separate projectors for the three colours, all focussed onto the same white screen. Something similar also happens in computer and television screens, with closely packed dots of colour. In each case, there is no interference between the colours—if the red light is projected, adding green or blue will not affect the red light itself, and the eye is free to see the mixture.

For printing on white paper, we have a different situation. Here we start with white light, but the printing ink filters some colours out—the more ink we add, the darker the result (this is called *subtractive* colours). For this purpose it is best to use not red/green/blue but the complementary colours, cyan/magenta/yellow. However, it is much more difficult to get the colours looking right. Most printers also use black ink (because overprinting the three primaries doesn't produce a good black); some do much more complicated adjustments.

For an artist, mixing coloured paints, the situation is different again. Mixing paints is closer to subtractive than to additive colours, but does not work exactly like subtractive printing ink. A more usual set of primaries for this purpose would be red/yellow/blue, but most artists use a much wider range of colours to mix.

Problems

The three-colour approach to images has proved successful, but it's worth exploring some of the issues around it.

First, let's think again about sodium light, and about taking photographs. If I photograph a sodium lamp, the three primary colour receptors in my camera will respond in a way which is similar to the response of the three types of cones. Then, if I display the resulting photograph on my computer screen (which uses LED technology), the image on the screen will be made up of a combination of red, green and blue LED cells. The challenge of displaying an image that looks good to me is the challenge of reproducing in my eyes roughly the same proportional responses that the original sodium lamp produced. The system might achieve this, though if we think in terms of the spectrum, it is very clear that the smear produced by my screen is hugely different from the monochrome sodium light itself.

Does this matter? Well, it might matter a lot.

For one thing, not all animal species are trichromatic as we are. Some have only two different colour receptors; some have four (specifically birds, reptiles and some fish). A tetrachromatic animal will see colour distinctions that we cannot see. Thus even if the screen image looks good to me, it would fail to satisfy the birds!

One interesting suggestion, not yet demonstrated, is that actually some humans have tetrachromacy—or at least that some of us have four different types of cones, which might give us effective tetrachromacy if we knew how to use them. I say 'us', but actually it's much more likely in women than in men, for genetic reasons. It may even be the case that some women are able to make use of them, and thus see a wider range of colours than most of us. But even if this does not happen, the responses of individuals may differ.

It is well known, of course, that some people are *less* sensitive to certain colour differences than the majority—this is normally referred to as 'colour blindness'. But if some people are more sensitive, or even if some people are differently sensitive, this means that something that I see as a good colour match might to these people seem a poor match.

Could there be a physical solution to this problem? Ideally, we might like to represent the full colour spectrum with many different finely graded colours. It would be possible to have more than three 'primary' colours, but it's very unlikely that we could go far in that direction with (say) cameras or display screens. Thus once again, what we have is a compromise.

Dots, lines, frames, pulses

Most discretisations of smooth variables in the world (but not all, as we have seen) involve dividing the continuity up into very many small steps. This works (when it does) because our perceptions do some of their own smoothing, and thereby restore some smoothness to something that is actually not at all smooth. This process probably involves not only the sense organs themselves, but also the neural processes that follow when sensory input is transmitted to the brain. In some cases, the sensory organ itself generates discrete signals even from smooth input, and these discrete signals must be interpreted smoothly. We have already seen how our eyes generate discrete signals for different colour ranges; it is also the case that brightness (or intensity) is conveyed in the eye-brain by the number of rods or cones that fire; in a given time interval, each one fires or does not, so at some level the internal process is digital anyway.

So some smoothing is natural, and this suggests that there is no problem about presenting data to the senses in discrete lumps, provided they are small enough. But it does raise the question of what 'small enough' means, and whether there are any other effects from such discretisation. The pointilliste artists like Seurat had a theory that their method of painting, building up colour shades from small dots of primary colours, actually enhanced our colour perception, making the images seem brighter.

Some recent films have been shot and played at 48 frames per second, rather than the usual 24. Although 24 fps is fast enough that the viewer is not normally aware of flickering, it seems that the smoothness of 48 fps causes some people to feel sick, from something like motion sickness. So there may indeed be effects of a rather oblique kind.

Three dimensions

Now let's return to 3D display, where we started this chapter.

A physical solution to 3D display would be to create a model image in 3D space, which one could walk around and see from different angles, just as much as if it were real. But this does not seem like a very good solution for (for example) 3D movies. It might work for scenes involving people in a room, but outdoor scenes with buildings would have to be greatly reduced in size, and those with distant vistas would not work at all.

The binocular method pioneered by Wheatstone is a much more plausible solution. As I have mentioned, it depends on the fact that a lot of our depth perception comes from our binocular vision, with the two eyes turning a little inwards to focus on something close. There are other effects—it's also the case that each eye does its own focussing—in something like the way a camera is focussed, by adjustment of the lens. However, this is really only important for very close objects. In the far distance, binocular vision doesn't help much either—one useful clue here on earth is what artists know as *tonal perspective*, where the intervening atmosphere causes distant objects to look hazier and slightly bluer than they would look close to (on the moon, with no atmosphere, it is impossible to tell how far away or how high the mountains are). And of course there is the usual kind of geometrical perspective—because we know what sort of heights humans typically have, one good clue to how far away they are is their perceived size.

In movies and still photographs, both kinds of perspective are present anyway, of course—and indeed, when watching a film, one is normally well aware of the three-dimensionality of the scene. The present 3D film technology does not attempt to adjust monocular focus, but does add the binocular vision component to enhance the illusion of three-dimensionality.

This illusion has some interesting components. For example, suppose you are watching a 3D film, and you see a post in the foreground and someone passing behind it at some distance. Someone watching the same film from across the other side of the room will see the same thing—the person and the post will line up with her eyes at the same instant that they line up with yours. This makes no sense geometrically!

Nevertheless, as with sound and colour, the illusion is what is important. A pragmatic mixture of physics and physiology may be quite sufficient to achieve a good illusion.

In the next chapter, I will consider other ways to represent particular kinds of images and sounds.

9. On perspective—and music

Apart from brief references to mosaics and weaving, I started Chapter 7 with the beginning of the mechanisation of image-making, the invention of photography. However, just as much as language itself (and probably for longer), the language of images has been developed over thousands of years of human history and prehistory. Our ability to understand, make sense of, interpret photographic images when they came along did not come out of thin air, nor was it entirely intuitive. It grew out of our earlier understanding, developed through art, of the possible relationships between the three-dimensional world and a two-dimensional representation of it. While I could not begin to chart all the ways in which painting and the other arts have contributed to the way we see photographs, there is one aspect that illustrates this contribution very well.

If you studied art at school, or perhaps later, you might have learnt about the rules of perspective (this might depend on your age, however!). This set of rules, this idea of a formal system of perspective, was an invention of the early Italian Renaissance. It is often presented as the 'correct' way to represent the three-dimensional world on a two-dimensional piece of paper. However, there are different ways of doing this, equally valid but with different characteristics. So before tackling perspective, I will look at two other domains.

Architecture

If you look at an architect's elevation drawing of a building, consider first the base of the building. It seems that you are looking at this with your eye at ground level. Now look at the roof: it appears now that your eye is at roof-level. You can see the same phenomenon at both sides, and indeed everywhere in the picture: your eye is always directly opposite the part you

https://doi.org/10.11647/OBP.0225.09

are looking at. An engineering drawing, of a machine part for example, will normally have the same characteristic.

This is sometimes known as parallel projection. Imagine a piece of paper held vertically beside the three-dimensional object. From each point of the object, imagine a line perpendicular to the paper; where it hits the paper is where that point of the object is represented. Of course in the case of a building, that would seem to require a piece of paper as large as the building! But such drawings are normally reduced in scale to a more manageable size—the important thing is that this rescaling happens after the parallel projection stage, when the image is already two-dimensional.

Architects and engineers employ such drawings for views of, say, the sides of a building based on a rectangular plan. They may also use them for angled views; to many people, such angled views look a little strange, because they do not follow the usual perspective rules. Parallel lines in the real world remain parallel in the drawing; while an artist working in perspective would expect parallel lines seen from an angle to appear to aim for a vanishing point. Figure 7 shows a cube in two different projections. You might find my perspective view slightly strange, because I have assumed that my sheet of glass is not vertical but angled down very slightly, so that even the vertical edges of the cube are not parallel in the projection. However, you would see such an effect as normal in a view looking up at a tall building, or (let's say in a horror movie) down a lift shaft.

The reason that parallel projection is common in architecture and engineering is that it has many advantages for those fields. In particular, measurements can be made on the drawing and translated, unambiguously and accurately, to measurements in the three-dimensional world. This is simply not true of perspective drawing.

Cartography

The problem with maps is not so much the three-dimensionality of the world (though that is interesting too) as the curvature of the earth's surface. One would like to represent any part of the world on a flat map, while preserving many of the properties of the real world in the representation—for example straight lines, direction and distance. For relatively small areas, this can be done accurately enough for most purposes. But for large areas

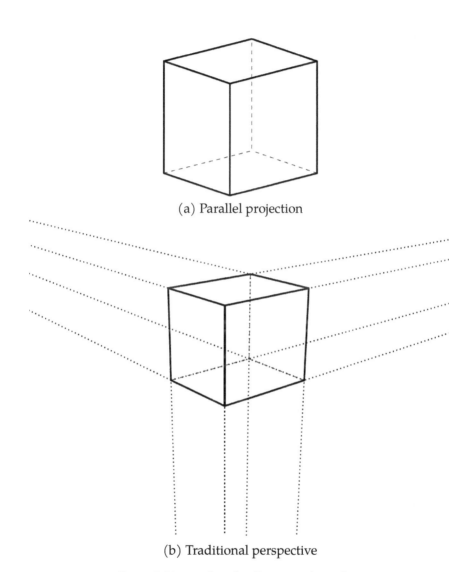

(a) Parallel projection

(b) Traditional perspective

Figure 7: Views of a cube. Diagram: the author.

(such as large countries, continents, oceans), it is impossible.

Once again, it is a question of projection—of projecting a part of the curved surface of the earth onto a flat plane. Cartographers have studied the question of what projection(s) to use since at least the second century CE (the Greeks, of course, already understood that the earth was round). One of the major reasons for such study is navigation. Many different projections have been proposed and used, and because it is impossible to preserve all the properties that one would like, each is a compromise.

Perhaps the best known is that of Mercator, which preserves bearings (that is, lines of constant bearing on the earth's surface are straight lines on Mercator), but makes for extreme variations of relative size. Figure 8 shows a map of the world in the Mercator projection (Google's map of the world is similar). Have a look at Greenland. Because Greenland is near the north pole, it looks huge on the Mercator projection. Now look at Australia, quite near the equator—it looks much smaller. Actually the land area of Australia is more than three times the land area of Greenland. The nearest equivalent to a straight line on the earth's surface is a *great circle* (a circle that divides the sphere exactly in half); in general, great circles look not at all straight on Mercator. And don't ask Google Maps to show you a map of the North Pole, or a proper map of the continent of Antarctica. The poles simply do not exist on the Mercator projection.

Figure 9 shows the world in the Peters projection. This projection distorts the shapes of the countries badly, but instead preserves their relative land areas.

Art

Have a look at a reproduction of a picture of a human figure taken from the wall of an ancient Egyptian tomb. Probably you will see the following. The head is in profile, seen from the height of the head. The feet are also in profile, seen from ground level. The torso is also seen from its own height, but full frontal. An example can be seen in Figure 10.

Now look instead at a human figure from a classical Greek vase (again, there is an example in Figure 11). Here again, probably each part of the body is seen from its own level, although without the changing view from front or side. Such a view seems to the modern eye much more 'realistic', but the projection is more akin to architectural projection than to perspective.

Figure 8: World map according to the Mercator projection
https://commons.wikimedia.org/w/index.php?curid=20066691
CC BY-SA 3.0.

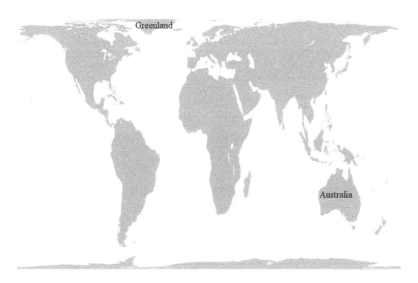

Figure 9: World map according to the Peters projection
https://commons.wikimedia.org/wiki/File:Peters_projection,_blank.svg
Public domain.

NEBT-ḤET.   ANUBIS.   SEB, or ḲEB.

Figure 10: Ancient Egyptian art: Drawings of the gods
Nebt-Het, Anubis, and Seb
https://commons.wikimedia.org/wiki/File:
Nebt-Het._AND_Anubis._AND_Seb,_or_Keb._(1902)_-_TIMEA.jpg
From Travelers in the Middle East Archive. CC BY 2.5.

Now look at a medieval picture with multiple figures. Very likely, the
central figure or figures are large, and peripheral or less important figures
are smaller. The relative location of the figures does not determine their
relative size; quite possibly there are foreground figures, nearer the artist
than the central figures, but smaller on the canvas. Figure 12 shows a small
section of the Bayeux Tapestry.

Now look at some paintings from the Edo period, 17th–19th centuries, of
Japanese art. Distant objects (like Mount Fuji) will be small, and there may
be closer perspective effects, but the foreground may well be represented,
from an angle, in something like architectural projection (see for example
the painting by Katsushika Hokusai, Figure 13). Now look at the paintings
of the Cubists in 20th century Europe. You may see a figure or object from
multiple angles simultaneously (see for example the portrait by Juan Gris
of Pablo Picasso, Figure 14).

All of these artists are choosing to represent the world in particular ways.

Figure 11: Drawing of an ancient Greek vase painting.
https://commons.wikimedia.org/wiki/File:
EB1911_Greek_Art_-_Vase_Drawing_(Fig._2).jpg
Encyclopædia Britannica (11th ed.), v. 12, 1911, p. 474—Public domain.

In the latter two cases, the artists concerned were probably well aware of the rules of perspective as laid down early in the Renaissance. But even in the earlier cases, the artists knew perfectly well that more distant objects look smaller: that if your friend walks away from you down the road, there comes a moment when you can cover her with your thumb stretched out at arm's length. You do not, of course, imagine that her size actually changes; nor do you assume for a moment that a person whom you first see at a distance is very small. Actually, with the help of other clues, you can probably judge fairly accurately the height of a person, whether he or she is right next to you or 100 yards away.

Some artists of the ancient world, particularly in classical Greece, thought seriously about how to represent 3D in 2D, and there are some pictures that reveal strong perspective aspects. But the real triumph of perspective had to wait until the Renaissance.

Figure 12: A banquet—detail from the Bayeux Tapestry
`https://commons.wikimedia.org/wiki/File:`
`Bayeuxtapeten,_Nordisk_familjebok.png`
Nordisk familjebok—Public domain.

Perspective in Western art

The artists of the early Italian Renaissance, such as Masaccio, Mantegna, and Piero della Francesca, wanted to create a variety of realistic art that would represent the world consistently, in some sense as we see it. They devised a set of rules for doing this—rather mathematical in nature. In fact Piero (a true Renaissance man!) was a mathematician as well as an artist, and wrote a mathematical treatise on the matter. The earlier diagram of a cube in perspective shows one aspect of this mathematical analysis: the way that any number of parallel lines in the real world appear in perspective to line up with a vanishing point—in the case of horizontal lines, this vanishing point is on the horizon.

We can think of the rules as follows. The artist places a flat sheet of glass vertically on a stand, and looks at the world through it. In order to do this properly, the artist has to fix the viewpoint—the point from which she looks through the glass—and not move while painting. Then she paints

Figure 13: Hokusai—Chushingura, Act XI, Scene 2
https://commons.wikimedia.org/wiki/File:HokusaiChushingura.jpg
Public domain.

on the glass, placing every part of the image exactly in line with the part of the world that she is representing. In other words, she projects the three-dimensional world onto the two-dimensional sheet of glass, using the (fixed) viewpoint as the point of projection. Figure 15 shows Albrecht Dürer's diagram of this process, showing how the artist has a fixed viewpoint.

The theory of perspective quickly became one of the staples of Western art. For several hundred years, artists were taught the rules of perspective, and by and large followed them. And the rest of us learnt to appreciate perspective, and to regard it as natural and, in some sense, the correct way to paint or draw.

The problem with perspective

Perspective is how the eye sees the world. And therefore painting a picture in perspective is the right way for the artist to convey to the viewers how she sees the world. Right?

Figure 14: Juan Gris—Portrait of Pablo Picasso
`https://commons.wikimedia.org/wiki/File:`
`Juan_Gris_-_Portrait_of_Pablo_Picasso_-_Google_Art_Project.jpg`
Public domain.

Well, in truth, there are interesting problems with perspective, and the
major one is revealed by the description above of how to do it. Remember
the fixed viewpoint? If the artist follows the rules, and if the viewer then
places his or her eyes in relation to the sheet of glass, *exactly* where the artist
was looking from when doing the painting, then the viewer will see what
the artist saw (or at least will see everything in the same geometric relationship). But if the viewer looks at the painting from *any other position*, the view
is distorted. Nonetheless, we have become so used to perspective pictures
that we no longer see the distortion.

Actually, even if they are painting in perspective, artists take some liber-
ties with the rules. For example, if the painting contains a full moon at the
top left of the canvas, an exact application of the rules requires the moon to
be an angled ellipse on the canvas, rather than a circle. Seen (only) from the
correct position, the ellipse looks like a circle. But because we typically do
not look from the right place, because we are also aware of the plane of the
painted surface as well as of the 3D world being portrayed, and because we
think we know what shape the moon is, the ellipse would tend to jar with

Q ii

Figure 15: Perspective drawing—From Albrecht Dürer's
Institutiones Geometricae (1532)
`http://www.cbi.umn.edu/hostedpublications/Tomash/`
Courtesy of the Erwin Tomash Library.

us. As a result, many artists would simply disobey the rules at this point, and give us a moon that is circular on the canvas.

Optical projection

Long before photography, it was known that a pinhole or lens could be used to project an image onto a screen. The pinhole idea was known in antiquity (since at least the fifth century BCE), and it was understoood that the image it produces is upside down because light travels in straight lines. In fact, although it is in reverse, with the reference point for projection in between the world and the image instead of the other side of a glass screen, the projection principle in the camera obscura is mathematically identical to that of Renaissance perspective. The first clear description of a camera obscura, a darkened room with a small hole in one wall, was by Leonardo da Vinci, but the principle was well known by then.

The principle of the magnifying glass lens was also known in antiquity, although lenses did not come into widespread use until around the 13th century CE, particularly with the invention of spectacles. Sometime in the 16th century, it was realised that a lens could be used instead of a pinhole in a camera obscura, gathering much more light to make a much better image. A lens and a mirror, mounted in a turret on top of a building, could be used to project an image (which could be viewed the right way up) into a darkened room below. Modern versions of this may be seen in various places, including the observatory at Greenwich.

A camera obscura with a simple lens follows the same projection principle, as does a photographic camera. Thus, when photographs came along in the 19th century, the images were instantly recognisable as using essentially the same kind of perspective that had been common in painting for several hundred years. Such images are indeed in 'correct' perspective, although the same qualification applies: they are only correct from a single viewpoint.

Camera lenses

Actually, most camera lenses are more complex than single magnifying glasses. In particular, many cameras have somewhat wide-angle lenses, while some have the opposite, telephoto lenses. Images created by such lenses are not in true perspective. We don't usually notice this—except in extreme cases. Such extreme cases include, in particular, very wide-angle or fish-eye lenses. We see the image produced by a fish-eye lens as distorted, simply because it does not follow perspective rules. In truth *all* 2D images of the 3D world are distortions—one cannot project 3D onto 2D without distorting it. But once again, we are so used to Renaissance perspective that we do not see it as distortion, even when we view it from the wrong position (which is almost always!).

Twenty-first century perspective

There is now, of course, a major industry in games for computers or games consoles. These games often present three-dimensional worlds, seen through the 2D computer screen, in which the player can move and operate. In general, games designers use exactly the same perspective rules that we are now so used to.

It is becoming more and more difficult to get away from the idea that these perspective rules give us the 'truth' about how we see the world around us. Renaissance perspective has become a universal in the language of images. The fact is that it is just as much an invention as writing—something that we had to learn about. One might argue that artists no longer need to learn how to do it—they can now leave it to the cameras. But all of us need to learn how to interpret perspective images. We do so very early in our lives, probably before we learn words, and certainly before we learn how to write. But it is a technological choice that we made, and not so very long ago.

Text, again

How do we see the relation between written and spoken language? In the several millennia since the invention of writing, our view of the status of written text, in comparison with spoken, has changed hugely. These changes occurred gradually, so that it is now difficult for us to reconstruct in our imagination what the relation might have been like in the distant past.

In classical Greece, for example, although they had a relatively well-developed writing system, they would not generally regard a book as itself an object of study. Rather it was an aide-memoire, to the writer or to someone else—an expert reader who would interpret the script and speak aloud, probably to an audience—in something like the way a modern musician interprets a musical score. We might also note that written material was much harder to read then; the ancient Greeks had little in the way of punctuation, and normally wrote without spaces between the words—as we have already seen in Chapter 5. The oral tradition remained the true source and method of propagating knowledge; writing played a strictly subservient role. Socrates, in particular, disdained writing, and all the written evidence that we have of the thought and work of Socrates was written by others, principally Plato.

The contrast with today could hardly be stronger. We know perfectly well that spoken and written languages are different, and play different roles, but we are entirely happy with the notion that a written document has its own validity, is to be understood and evaluated on its own terms. This statement applies not only to books, but to magazines, pamphlets, letters, emails, text messages, posters, notes, and so on. In many cases, we regard the written form as pre-eminent—for example, in laws and contracts. In ancient Greece, a contract was oral, requiring witnesses who could attest to it orally—any written form was simply a reminder, with no legal or contractual status.

If you were to come to one of my talks, and I were to attempt to convey the same or similar information to you orally, as I am currently trying to do through text, I would do it quite differently. But the fact is that this is unlikely to happen. This book will have to stand on its own as my message to you: I have to get everything I want to say to you into this written text.

Coding and digital text

When Morse devised a scheme for transmitting messages over electrical wires, he probably thought of his codes as instructions. That is, the sending operator is given a message, let's assume on paper, and the first letter is 'A'. The operator turns this into dot-dash, and transmits this signal. The receiving operator hears it, and writes down 'A', as instructed. The dot-dash is merely a way of getting the 'A' from one piece of paper to another.

A similar attitude is evident when the ASCII coding scheme was invented in the twentieth century, with the human operators replaced by machines. As we discovered in Chapter 5, ASCII codes include both codes for letters and codes for machine operations. ASCII code number 65, which is 1000001, says to the machine at the other end "print an 'A'", while code number 13, which is 0001101, says "return to the beginning of the line".

However, we are moving away from this notion. One view, which is probably common now, is that the coded form of a text (inside a machine or traversing a wire) is as valid a representation of the true text as is a printed document. Another view, which may seem extreme but has some advantages, is that the coded form is the true document, while a printed form is just a representation.

Why would that have advantages? Well, while such things as formatting (fonts, layout etc.) seem to be intrinsic to a printed or displayed text, these things are not normally seen as part of the essence of a text. We are usually entirely happy with the notion that, if a document is printed on different sizes of paper, or in a larger font for people with poor eyesight, such things as the line breaks in each paragraph and the spread of text over successive pages may be adjusted to fit, without altering the essence of the text. The same applies to a screen display in a window—and indeed it is very annoying when the text in a webpage is set up (pre-formatted) with a fixed width that is wider than the window you are trying to read it in. Thus the unformatted digital file seems to capture the essentials of the text, while the printed or displayed version mixes up the essence with more superficial aspects.

All this is quite aside from all the advantages of manipulation and processing that can follow from having a coded text, such as cutting and pasting, or searching for words or phrases, or counting words.

These days, it is quite possible for a page of text to be represented in a machine as a (scanned) image, rather than as coded text. But this representation has all the limitations of the printed page itself. The true text now seems to reside in the coded version, not in the scanned image, nor on paper.

Images

Thinking about photographs, as we have seen in Chapter 7, we seem to have made a similar transition. Within a period of about 30 years, from the mid-1980s to 2010, we came to accept that a 'photograph' *is* a digital file. Forty years ago, if you had said "send me a photograph", I would have interpreted that to mean a chemically-produced print on paper, to be sent by old-fashioned post. Now (it does not even occur to me to question this), you mean a digital file sent by email or MMS or placed in the cloud for you to download. The digital file *is* the photograph—how you choose to display it is another matter.

But this is not true of all images. Artists still paint pictures onto canvases, and in such cases the picture is the physical object. If we subsequently photograph it, the photograph (in its digital form or printed or displayed) is an attempt to represent the painting, which we may regard as more or less successful, but at best a partial substitute for the real object.

It's true, of course, that some artists (for example David Hockney), and most designers, now create directly in digital form. Hockney's wonderful iPad art is created directly on the tablet, and once again the digital file *is* the picture. Nevertheless, both the activity of painting and the resulting art objects retain their status outside the digital world.

Music

In a similar fashion, there are alternative ways to represent music. Of course these have existed for a very long time: musical notations can be found on cuneiform clay tablets from around 2000 BCE, and the ancient Greeks had a more developed notation from around the 6th century BCE.

The stave notation with which we are familiar today developed gradually over a number of centuries, from about the 11th century CE. Its ability to indicate pitch and duration of individual notes, their relative timing, and the overall rhythm came at different times. It is most strongly associated with western classical music, though it is also used in other musical genres and contexts.

Here the question of how it relates to the real musical sound is quite complex. At one level, we see it again as a set of instructions to a singer or instrumentalist—in order to make this music, *this* is what you have to do. But we expect a player to *interpret* the music—and therefore the sound that this player produces is not necessarily the same as another player of the same music. Indeed, we find the differences important and interesting. This view is actually not far from the ancient Greek view of written text—it was there for a human reader, reading out loud, to interpret. In most musical environments outside of classical music, the sound is developed and learnt through performance, and written musical notation, if used at all, is treated as an aide-memoire.

Nevertheless, in western classical music, the power of this abstraction of sound to musical notation might be seen to take us a little nearer to the modern view of text, if not quite there. In the case of a late work by Beethoven, for example, we might almost see the score as being the real music, and a performance of it as a (maybe flawed) 'display'. The knowledge that Beethoven was deaf when he composed it, so that the paper version is the only indication we have of what he heard internally, reinforces this view.

Digital graphics and music

In the computer age, the kind of pixel-based image associated with photographs and scanned images is not the only way in which graphical information can be represented in digital form. For example, there is an alternative that is commonly used in design and engineering applications, where the designer creates graphics directly on a computer, or where the computer itself builds the image from something more abstract still, like a three-dimensional model. This is known as *vector graphics*—the basic idea is of describing the shapes to be displayed. For example, the digital representation might specify that there is a point *here*, and another *here*, and they are joined by a straight line of this colour and thickness.

Vector graphics does not predate computers, but was developed in the very early days of computing.

Similarly, we have different ways of representing music in digital files. Just as vector graphics attempts a more abstract representation of certain types of image than using pixels, we have ways of representing some kinds of music that are more abstract than the pulse-code-modulation representation of sound. The best-known example is the MIDI system. MIDI was developed in the 1980s, well into the computer era, to allow digital control of the playing of instruments. It allows the specification of notes to be played, including pitch, onset and duration, all of which, of course, are represented in the traditional stave notation of a musical score. But it can also include further elements, such as envelope (the way the note fades or develops over time).

Thus within the computer era, there has been a proliferation of ways to represent and manipulate different kinds of text, still and moving images, and sound. Part of the point of such representations is to extend the possibilities for the computer manipulation of data. Thus, for example, vector graphics allows a relatively simple computer operation to extend a line or move its endpoint—much more difficult if the line is represented only by its pixels. Such computer manipulation is beyond the scope of this book. However, in the next three chapters, we will consider further kinds of data where the idea of machine manipulation predates computers.

10. Calculation

Let us now return to numbers.

After the invention of the zero and positional notation by Hindu mathematicians, and the systematisation of the rules of arithmetic by the Arabs, it was only a matter of time before we would turn our attention to calculating by machine.

Machines that calculate?

In one sense, a kind of mechanically-aided calculation had existed for far longer, in the form of the abacus. In fact, the abacus involves a kind of version of the positional notation of the later Arabic system—a column or row of the abacus is still there, even if it has no beads in it, so the zero is represented. And the Arabic rules were embodied in the knowledge of trained abacus-users long before they were codified.

The abacus helps the human to calculate (and also to remember the final result of a calculation)—it does not do the calculation itself. The Hindu-Arabic system (both the representation of numbers and the rules of arithmetic) allowed the use of pen and paper with the same speed and accuracy as the abacus, with the added advantage of retaining a permanent record of the calculation. Machines that calculate? That would be revolutionary, but this system also gives us the foundations of such a revolution. Given that each column of a number is to be treated in exactly the same way as all the others, and the rules have been codified, mechanisation is invited.

One person who first made serious inroads into this idea was Blaise Pascal. In the first half of the seventeenth century, as a precocious teenager, Pascal developed a calculating machine to help in his father's financial calculations. Numbers were dialled on a series of wheels. The machine could be used to add, and, by a process of making subtraction look like addition,

© Stephen Robertson, CC BY 4.0   https://doi.org/10.11647/OBP.0225.10

also to subtract. It could not yet multiply or divide. A similar machine, which has not survived, was invented by Wilhelm Schickard about the time of Pascal's birth, and another famous mathematician, Leibniz, devised another such machine shortly after Pascal.

Logarithms and the slide rule

Pascal's and Leibniz's machines represent, in direct mechanical form, the Arabic rules of arithmetic, whereby numbers are made up of discrete digits arranged in columns. This makes them the forerunners of the digital electronic calculators of the late 20th century.

But this is not the only approach to mechanical aids to calculation. A very different method takes what might be described as an analogue approach to calculation (as before, we can regard analogue and digital as opposing principles). This was happening in parallel with the work of Shickard, Pascal and Leibniz. It is clear that the time of mechanical calculation had arrived, even if the methods were still in dispute.

The principle of the logarithm was developed by Napier, around the beginning of the seventeenth century. The characteristic of logarithms is that they convert multiplication into addition. That is, in order to multiply two numbers, you take their *logs* (that is, you look up in a table the logarithm of each number). Then you add them together, and then you take the *antilog* of the result (that is, do the reverse lookup in the table). Great efforts were made to compile accurate log tables over the next two or three centuries.

An alternative is to have the numbers represented on a scale (like a ruler or measuring tape—see Figure 16), but spaced according to their logarithms rather than in the usual equal spacing. Then two numbers can be multiplied by adding their lengths on this scale. Initially this was done by having a single scale and using dividers, but subsequently the idea of two scales that could be slid against each other replaced the single scale. The slide rule was born—see Figure 16.

In setting up a multiplication on a slide rule, you move B scale so that the 1 (on B) lines up with one of the numbers you want to multiply (on A). Then you look up the second number on B and read off the result from the corresponding position on A. So a number is a position; you can set it more or less accurately (just as you can measure with a ruler more or less

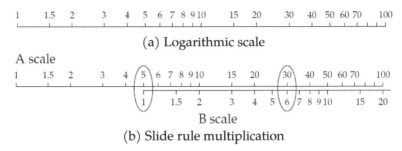

(a) Logarithmic scale

(b) Slide rule multiplication

Figure 16: Slide rule. Diagram: the author.

accurately). The digits that you might use to write down the number are not involved. This is why we might describe it as analogue; the position is an analogue representation of the number, not a digital one.

Provided that some degree of approximation was acceptable, the slide rule principle was an effective method for multiplication and division for three centuries or so before digit-based calculators could compete in this domain, and remained in use for most of another century. But in the late-twentieth-century IT revolution, when digital principles ran riot over vast regions of human endeavour, the slide rule lost its status as the pre-eminent method of calculation favoured by scientists and others.

However, I jump ahead. First, we have to see how the digital triumph began.

The comptometer

Many inventors (or would-be inventors) of mechanical calculators, from the mid-nineteenth century on, had in mind the idea of a key-driven device. That is, one would key in the number on some form of keyboard. This looks like a different principle from the Pascal system of setting dials. However, it shares with Shickard, Pascal and Leibniz a digital view of calculation.

This idea took some time to become a reality. Essentially, the difficulty lay in precisely those rules of arithmetic that the Arabic mathematicians had given us, and which had inspired the quest for mechanical calculators in the first place. The particular rule that causes difficulty in the design of key-driven calculators is the carry rule: carrying from one column to the next. Part of the problem is the recursive nature of carry: a carry to a second column might trigger a carry to a third, and so on. (Pascal had, and resolved, the same problem with his wheel-based calculator.)

It is interesting to speculate here, as in the discussion of the typewriter, why keys were regarded as so important. We have of course the very considerable history of keyboard musical instruments, and later, while the key-driven calculator was still struggling, we have the successful development of the Hughes telex-like machine and later the typewriter. It seems that something about the keyboard as a transparent method of controlling a mechanical device really appealed to Victorian inventors—a sort of Platonic ideal of fingertip control which one can see reflected in many 20th and 21st century approaches to design.

At any rate, the mechanical problem was eventually solved, and by about 1890, there were full-function key-driven *comptometers* for standard arithmetic operations—'full-function' implies that they could be used to multiply and divide as well as to add or subtract. In fairly short order, there were also printing calculators, and then, in the twentieth century, electrically powered and eventually electronic devices.

Babbage

A different view of mechanical calculation is due to Charles Babbage, the nineteenth-century mathematician and inventor (active from the 1820s until his death in 1871). His name is associated with two machines that are seen as the ancestors of modern computers, the Difference Engine and the Analytical Engine. Actually, he failed to produce a working version of either machine, although forms of the Difference Engine were made subsequently. The design of the much more ambitious Analytical Engine anticipated modern computers in many interesting ways.

These machines, like comptometers, can be described as digital (the representation of numbers used decimal digits rather than the binary form common today). The Difference Engine was a calculator on a grand scale, but aiming at automating the repetition of many similar calculations, rather than just a single calculation at a time. Thus it can be seen as anticipating the idea of *programming* a machine to do many calculations. The Analytical Engine would have been programmable to undertake many different kinds of tasks, by implication going well beyond what was seen as arithmetic calculation.

The ostensible purpose for developing these engines was an odd throwback. Babbage did not imagine them, or at least did not sell them to his backers, as serving general calculation purposes. Rather they would be used specifically to calculate the tables—tables of logarithms and the like—that would be published and distributed, in the already established way, to allow people to do their own, unmechanised, calculations and computations. A major reason for the work was the known fact that existing tables contained many errors, primarily because of the involvement of people doing tedious repetitive tasks at all stages of their construction.

Input and output

This aim, generating printed tables, led Babbage to think about other issues than calculation, in particular the input and output stages. For output, he sought to automate part of the printing process: in particular, to have the machine construct the plates for printing, thus avoiding typesetting errors. (Plates for printing, as opposed to movable type, have a long history, briefly discussed in Chapter 4.) This anticipated by more than a century the revolutionary effect of the computer on the printing industry.

He also addressed the question of input—both of numbers and (in the case of the Analytical Engine) instructions to the machine as to what to do. The notion of keyboard input, so important to the comptometer inventors, was of no interest to him. Instead, he proposed making use of Jacquard's invention of punched cards, which we encountered in an earlier chapter. Jacquard successfully used punched cards to control looms; Babbage would set them the task of controlling his Analytical Engine. In the event, the punched card idea would be taken up at the end of the nineteenth century for another purpose altogether, as we shall see in the next chapter. But eventually, in the 1950s, they would be used in much the way envisaged by Babbage.

Computability

Babbage and his collaborator Ada Lovelace began to develop some general notions of what might be 'computable', in other words, what kinds of task might be susceptible to being delegated to a machine. Arithmetic calculation was clearly in this category, having been reduced to sets of rule-driven steps ("algorithms") by the tenth-century mathematicians of the House of Wisdom in Baghdad. But Babbage and Lovelace believed that the possibilities went far beyond calculation.

Although a working Difference Engine was eventually constructed, Babbage's ideas mostly died with him. The twentieth-century inventors who brought the modern computer into being were largely unaware of Babbage's work.

Nevertheless, the notion of *computability*, what might in principle be computable, would be taken up by twentieth-century mathematicians such as Kurt Gödel, Alonzo Church and Alan Turing. In the 1930s, Alan Turing described (as a mathematical abstraction) a general-purpose computing machine—and subsequently made major contributions to the code-breaking effort in the Second World War which, as we shall see in Chapter 12, led on to the development of actual computers.

But something else happened in the 1890s, which brought machines into a rather different form of work in a very practical way. This is the subject of the next chapter.

On numbers and machines

Depending on your age, you (or perhaps your children) may well have learnt at school about binary numbers and binary arithmetic. Using basically the same positional notation system as bequeathed to us by the Arabs, but only two digits, 0 and 1, we can represent any numbers and perform any arithmetical operations. In this scheme, the rightmost digit represents the units, but the next to the left represents the twos rather than the tens, and the next after that the fours ($2^2 = 4$) rather than the hundreds.

You will probably also know that computers operate with bits (bit is an abreviation of *bi*nary dig*it*), and you may well associate the digits of binary arithmetic with the bits in the computer. Thus inside computers, numbers are in binary, because that's what computers know about and work with, right?

Well, no, not quite. Actually, it is entirely possible to hold all numbers and do all arithmetic inside a machine in the traditional decimal Arabic system. The system is called BCD (binary coded decimal); it involves a direct representation of each of the decimal digits, and rules for arithmetic operations on them that would be familiar to a primary-school child. Furthermore, retaining the conventional decimal structure has some advantages over converting to pure binary—for one thing, it is difficult to approximate a number according to the rules usually applied to decimal numbers, if you are operating in the pure binary representation. Babbage's machines were to have worked in something akin to BCD; modern electronic calculators usually work in BCD.

In fact, there is more than one pure binary version. There is a form of representation, slightly different from the one described above though using essentially the same arithmetic, called *two's complement*. This has some advantages, including simplifying dealing with negative numbers, and is often used in computers. Thus we have at least three different forms of representation and two different forms of arithmetic. There are also different ways to represent fractional numbers, very large or very small numbers, and numbers that require great accuracy.

As a general rule, machines can convert numbers from one representation to another, internally, and back again. Thus we do not see all these various representations or methods—we just see the results in the usual decimal system.

11. Data processing

In the United States, a census of the population is undertaken every decade. The 1880 census took 8 years to analyse fully, and the Census Office badly wanted to be able to complete the analysis of the 1890 census in a very much shorter time. The solution to this problem was developed by Herman Hollerith.

Tabulation

Every analysis of the census data involved sorting returns into categories, counting the number of returns in each category, and recording the resulting numbers. Any of the answers to the census questions, singly or in combination, might be the basis for categorisation. This process of sorting and counting can be described as *tabulation*, the creation of (printed) tables.

Hollerith's system involved punched cards—using ideas from Jacquard, whom we encountered in Chapter 7 (he probably did not know Babbage's work). Each individual census record (representing a person) would be encoded onto a punched card. Each census question would be associated with a group of punch positions, and the individual answer encoded as holes punched in some of these positions. Thus for example the age question had positions for different age-bands. For the gender question there were two positions, M and F, and a hole was punched in one of them. You might think that this datum needs only one position, where the absence or presence of a hole would indicate male or female. But this would assume (a) that this question is always answered one way or the other, and (b, more seriously for the Hollerith system) that the counting mechanism is just as able to count absence as presence of holes. Well, alternatively you could do something involving subtracting numbers in the final table. At any rate, Hollerith did use two positions for this and other yes-no questions, despite

    https://doi.org/10.11647/OBP.0225.11

the limited number of hole positions at his disposal.

Hollerith developed a machine called a tabulator, into which an operator would insert each card in turn. The machine would automatically operate a number of counters, incrementing them as appropriate for each card. A counter could be associated with a single punch position, thus recording the number of cards with holes in that position—for example, a counter for each age range. But the machine also allowed questions to be cross-tabulated—counters could be set up for combinations of holes. In the first version of the machine, these combinations were pre-set, but subsequently Hollerith developed a plugboard that allowed new combinations to be set up. This effectively made the machine, at least to some degree, programmable.

The operator places each card in turn in the machine, and the appropriate counters (dials) are incremented, according to the holes punched in the card.

Figure 17: Hollerith tabulator,
From *Scientific American*, 1890.
http://www.cbi.umn.edu/hostedpublications/Tomash/
Courtesy of the Erwin Tomash Library.

The cards that Hollerith used for this purpose were the same size as the then US banknotes. This was a deliberate choice by Hollerith, partly because it enabled him to get cheap boxes for his cards! Although the size of US banknotes has been reduced since then, the card size remained the same, and is enshrined in an international standard.

The uses of tabulation

Hollerith's invention allowed the 1890 US census analysis to be completed in two years, and under budget. It was very quickly taken up for many other national censuses. But it also became clear that it could be used for many other purposes, allowing the analysis of different kinds of data in ways that would not have been feasible when everything was done manually. The invention begat an industry, which became known as the tabulator industry; the analysis of data using tabulators and all the associated machinery became known as *data processing*.

The system was adopted by insurance companies and others. Already in the mid-1890s, one of the US railroad companies used the Hollerith system to compile statistics relating to freight traffic, and to audit the accounts. Other businesses followed suit. The range of applications included account transactions, payroll accounting, inventory management and billing.

For over half a century, tabulator-based systems were used for a huge variety of data-processing purposes, by a huge variety of companies and organisations. When I started working in the field of information retrieval at the end of the 1960s, there were still companies using tabulator systems for searching scientific databases (in something akin to the way we search the web using a search engine such as Google)—though inevitably by that time they were being replaced by computers.

In the meantime, punched cards had developed a new lease of life as an input mechanism for computers—as we shall see below.

Unit records

Back in Chapter 6, we discussed the idea of a *database*. A collection of data of the kind for which tabulators were used, such as a set of census returns, is a kind of database. It's worth exploring its character in database terms.

In my description above of census data, I talked about records relating to each individual: one punched card represents one person. In this case, we might consider it to be a *flat* database, that is one with just a single type of record. However, census data is typically a little more complex than that. In particular, data is normally (at least currently in the UK) actually collected by household, rather than by individual. A single household with a single address quite likely includes more than one individual.

In a modern computer database, the organisation of the data would probably reflect this, by having each individual record linked to a separate household record. So the database would have (at least) two different types of record, together with a linking mechanism. The household record would contain data such as address, and the individual record would give age, gender, etc. Whether or not census data is actually analysed in this way, it is easy to imagine questions one might ask that cut across these record types: for example, 'tabulate households by the number of school-age children in each'.

A flat database of unit records of a single type, such as that built on Hollerith punched cards for the 1890 US census, may serve very well for certain kinds of analysis but does have limitations. It would certainly be possible to derive two different sets of unit records from the same basic data: for example, by punching a card representing each household—and keeping these separate from the individual cards. But the linking would be difficult, and some analyses would not be possible.

All the many applications of tabulator technology, from the invention of the Hollerith system until they died out in the 1960s, had to be approached with these limitations in mind. Thus each depended on a choice of a unit as the basis for the unit records. Separate sets of unit records might be created for different kinds of unit, to be analysed separately, but not to be linked subsequently, except by hand.

The business

Despite such limitations, tabulator systems were adopted widely. Hollerith formed a company to exploit his invention. Hollerith's company merged with three others to form CTR in 1911; in 1924, the merged company was renamed as International Business Machines—IBM. In the fifties and sixties, IBM redefined itself as a computer company; in the 80s it developed the PC (not the first personal computer, but one of the most successful commercially). But before that, still in the first quarter of the century, other rival tabulator companies were formed, and developed in other ways. For example James Powers, who had taken over the construction of census-analysis equipment within the US Census Bureau, formed the Powers Tabulating Machine company in 1911. This eventually merged with the typewriter company Remington, which we have already encountered, as Remington Rand.

The development of new business and commercial rivalry went hand-in-hand. JoAnne Yates, in the article *Co-Evolution of Information Processing Technology and Use*, has a fascinating discussion of the relationship between the insurance industry and the tabulator companies. Many technical innovations were driven by the demands of the insurance companies, who tended to play off the rival tabulator companies against each other, in order to get the developments they needed.

Some of these particular innovations are:

- the ability to accumulate (sum) numbers encoded on cards, as well as simply counting them (this eventually evolved into full-function arithmetic);

- the ability to sort a deck of cards into categories (Hollerith's tabulator included a rather simple aid to human sorting, but the insurance companies wanted a much faster process);

- the ability to print, generating reports automatically (Hollerith's design required a human operator to transcribe the numbers indicated by the counters); and

- the ability to encode, and therefore print, alphabetic characters (e.g. names and addresses) as well as numeric ones.

In the third case, for instance, the Powers company had a working printing device considerably before IBM. However, IBM would eventually catch up with a good product, and its considerably greater commercial muscle enabled it to dominate the industry for many years. The last case was initiated when one of the UK insurance companies took over the UK arm of the Powers company.

Naturally enough, the first (and virtually all subsequent) alphanumeric keypunches used the Sholes QWERTY keyboard with which we are already familiar. And indeed, in 1928, IBM bought up one of the earliest electric typewriter companies, Electromatic, and became a major player in the type-writer business, eventually coming to dominate the market in high-end of-fice electric typewriters, which effectively carried it over the transition to computers.

The transition

The continuity between the tabulator-based data-processing world and the computing world is remarkably strong. For one thing, at least as far as IBM computers were concerned, punched cards were used for data input, both for the data to be analysed and for program instructions. This meant that the keypunches and the card feeders and the card readers, as well as the cards themselves, could be repurposed, and trained keypunch operators could be reassigned.

But even outside the IBM environment, the conception of the business uses to which a computer might be put started from what data processing had done before computers came along. Of course companies innovate, de-vise new ways of using the resources that they have, including computers. But these innovations seldom come out of thin air, and often involve do-ing something that you are doing already, but in a better or more flexible or more reliable way. And besides, those functions that are most suscepti-ble to mechanisation using tabulators are also those functions that are most obviously computerisable.

In the late 1940s, after the end of Second World War, academics took up the idea of developing computers—some sense of what might be possible having survived the total destruction of Bletchley Park (see the next chap-ter). They of course had many ideas about what might be done with com-puters, to what uses they might be put—many of them revolving around cal-culation, since the people who worked on the computers were often math-ematicians, physicists and engineers. Applications included astronomical calculations, weather forecasting, mathematical modelling of engineering structures or chemical processes, and so on.

But as a business, operating in the world of business, computing could only take off by starting somewhere that other businesses might recognise as useful. That starting place was the existing tabulator industry and its existing applications. As we have seen, these applications were not primarily about calculation. They were about sorting, storing, manipulating and managing records.

Data

I have talked about databases, but it is worth thinking a little more about data. In the Jacquard loom, a hole in a card represents a control mechanism for a machine, and it is really only in hindsight that it looks like a coded representation of the woven pattern. Babbage's notion (which of course was never made concrete) was that such holes could encode not only numerical data (to be operated upon), but also the arithmetic operations that might be applied to them, and indeed therefore the sequence of operations and the rules for branching, which we would now call the program.

Such abstract considerations were probably far from Hollerith's mind when he developed his system, but he did have a notion that they could encode characteristics of people. One of his sources of inspiration was a system in use on the railways, where the conductor would punch each ticket with codes to identify the passenger. Another source was the piano roll, a continuous roll of paper in which slots were cut to represent and reproduce notes played on the piano, an almost-digital representation of music (it's not properly digital—the length of each slot is an analogue representation of the time for which each note is held). Thus there is the beginning of an idea that there is some kind of abstraction that we might call *data*.

Over the next half century or more, this notion will develop slowly. In particular, the idea is centred around the association with characters that may be typed or printed. And we know about characters—we have letters and digits. Oh, and special symbols and punctuation marks. Oh, and we discover that we need to extend the notion a little, to include first spaces, then carriage returns and line feeds (the ability to print a three-line address from an IBM card came really late in the day). In fact, basically anything that can be a key on a typewriter—as typewriters were electrified, they were also given a key for carriage return, as opposed to the lever on the left-hand-end of the carriage, which was the norm for manual typewriters.

Despite Morse and more than a century of telecommunications, it was not until the 1960s, well into the computing era, that the idea would emerge that we might need a standard form of encoding for all these characters, for all the different data processing and telecommunications purposes. Then two different standards appeared: ASCII, which was explicitly designed to be a standard, and EBCDIC, which became a *de facto* standard because of its use by IBM (we encountered these in Chapter 5). Both standards take as their basis the kinds of characters that one sees on a typewriter keyboard.

As with numbers, the internal storage and manipulation of such data in computers does not necessarily retain the direct link to printable characters. We have already seen that numbers may be converted and then stored and manipulated in a number of different ways—the same applies to other kinds of data. This was already true of the original Hollerith card—the gender question is represented by two hole-positions called M and F, rather than by the words Male and Female. But many other such transformations may be made inside a computer, even if the external representation (input from a keyboard, or output to a printer or a display) is always made using the words with which we are familiar.

Now, finally, we can think of all text and all numbers as data, and in particular as digital data, storable, transmittable and processable in a large variety of ways, but essentially all by the same kinds of machines. It will take a little longer for the idea of data to encompass images and music as well, but that will happen eventually, as we saw in Chapter 7. Furthermore, not only *can* we think of text and numbers and photographs as data, in some sense we *have to* think of them this way. As with so many other influences of technology, there is no going back.

12. Ciphers

It is easy to think of computers as giant calculators, and indeed the task of calculation and its mechanisation contributed both to the idea of constructing such a machine and to the conception of the tasks to which it might be addressed. Difficult calculation tasks such as those involved in ballistics (particularly in wartime) provided some of the stimulus towards the post-Second-World-War development of computers. But we have just seen, in the previous chapter, how a really rather different kind of task stimulated a form of mechanisation that brought us close to the computer era. The major stimulus for the actual invention of computers came from another domain again.

The challenge was that of breaking the codes used by enemies in order to be able to read their supposedly secret messages, in the technological hothouse that was the Second World War. In order to see how this came about, we again start much earlier.

Codes and ciphers

Throughout history, people have felt the need to write messages (point-to-point messages, in terms of our previous discussion) that would be unreadable to anyone other than the intended recipient, specifically to anyone who might intercept it en route. Military commands, intelligence reports, instructions to agents, love letters, arrangements for meetings, plans for any kind of action or activity that could prompt counter-measures of any kind by any third party—all these and many more might be deemed by the sender to need encryption.

Since the word *code* is somewhat overloaded in present-day usage, I will use the word *encryption* to indicate putting some message into code, in such a way that it can only be read by someone who has the key to the code, and

          https://doi.org/10.11647/OBP.0225.12

cipher for the method or rules for doing so. The original message is *plain* text and encryption results in the *encrypted* or *cipher* message. Recovering the plain text (given the key) is *decryption*. Discovering the key, or even the complete cipher system I may still refer to as *code-breaking*, in deference to popular usage. The whole subject, of designing ciphers and of breaking them, and of studying their properties (such as whether in principle they are breakable) is *cryptography*.

A book with a marvellous account of the different kinds of ciphers that have been used through history, and of the efforts of opponents to break them, is Simon Singh's *The Code Book*. Much of the rest of this chapter is drawn from Singh's book.

The alphabet and encryption

From the beginning and to this day there has been some use of word-based coding systems. A report in a newspaper on my table today describes a case in which some alleged terrorist plotters "used code words" for some possibly suspicious-sounding words, like firearms. But such systems are really intended to disguise or camouflage a coded message, rendering it less suspicious and therefore less likely to attract attention. Another approach is to hide the existence of a message altogether.

However, most of cryptography addresses the question of how to render a message unreadable even when the adversary is in possession of what he or she suspects or knows to be a cipher message. Once again, it is hard to conceive of much of the history of encryption without the alphabet. Most encryption systems throughout history have been alphabet-based. Ciphers typically involve either or both of: re-arranging the letters of the message, and/or substituting different characters for those in the message. Even in Japan and China, we see evidence of the use of alphabets or alphabet-like symbol sets for encryption. Japanese ciphers tend to be based on one of the phonetic alphabets (*kana*), while a Chinese cipher might use, for example, either a phonetic alphabet or the so-called Four Corner method of encoding each character into four or five numbers, which is also used as a sort of substitute for alphbetical order, for sorting and then looking up characters.

Given an alphabet, one of the simplest kinds of encryption is to substitute for each letter in a message the letter three places further on in the alphabet (this was a cipher used by Julius Caesar). If I do this with the heading of this section, I get

Wkh doskdehw dqg hqfubswlrq.

Or I could choose a different shift, or I could rearrange the cipher alphabet in some way. My intended recipient needs to know what cipher system I have used, a key that will enable him or her to decrypt the message: both the principle ('alphabet shift', for example) and the number of characters shifted.

But as we shall see in a minute, such ciphers, in which a plaintext *e* is always represented by the same symbol in the cipher message (in this case an *h*) are normally very easy to break. To make a stronger cipher, we might use all 26 possible shifts of the alphabet, and a key that tells us which shift to use for which letter. The key is a word, whose first letter tells us which shift to use for the first letter of the message, second for the second, and so on. When we reach the end of the codeword, we return to the beginning. This is the basis for a Vigenère cipher, invented by Blaise de Vigenère in the sixteenth century.

The Vigenère cipher makes use of the Vigenère square, showing all possible shifts of the alphabet (see Figure 18). Suppose that we again want to encrypt the heading of this section, and the codeword is *revolution*. We write out the plain text, and underneath it the codeword, repeated as many times as are necessary to match every letter of the plain text. Then we look up each plain text letter in the top row of the Vigenère square, and encrypt it with the corresponding letter in the row identified by the codeword letter. The first lookup (column T row R) is circled in the figure. Given the codeword, decryption is equally simple—but you need the codeword.

```
THE ALPHABET AND ENCRYPTION   plain text
REV OLUTIONR EVO LUTIONREVO   repeated codeword
KLZ OWJAIPRK EIR PHVZMCKMJB   cipher text
```

The Vigenère cipher is much stronger than the simple substitution of the alphabet shift, and was thought to be unbreakable. In the example, you can see that the two *A*s in *alphabet* are represented by different letters in the cipher text. But it can be broken—the man who established this fact is

Figure 18: Vigenère square. Diagram: the author.

someone we have already encountered in Chapter 10: the nineteenth century mathematician and inventor Charles Babbage.

Code breaking

Suppose that I am in possession of a cipher message, or a set of such messages from a single source—but that I am not the intended recipient, and
do not know the cipher. If I have any reason to believe that the cipher is a
simple alphabet shift, or indeed any simple one-for-one substitution, then it
should be easy for me to discover the key and thus decrypt it. In particular,
the number of occurrences of each letter will provide a clear clue as to which
letters might have been substituted for, say, E or T or A (the most common
letters in English). The longer the message the easier this is, but in the above
short message I have three each of E and T, and also N, and only two As

Even so, we see immediately that breaking is a different kind of task from encryption and decryption. Encryption, and decryption for the recipient in possession of the key, both involve following a very simple set of rules. Breaking the cipher, however, is a little more complex. The code-breaker may have to do some counting and statistics, and then try out a number of possibilities.

Babbage's method of breaking the Vigenère cipher involves looking for repeated sequences of characters in the cipher message. The distances between such sequences will give good clues as to the length of the keyword used, after which an extended form of the analysis of the statistics of letter occurrence, as used to break simple substitution ciphers, is likely to be effective.

However, using a longer key (for example a phrase or an entire poem) makes it more difficult to break. The final stage of this development was to construct a whole series of long random keys, each printed on a separate sheet of paper, forming a pad, of which sender and receiver would each have a copy. The sender would encrypt a message using the first sheet, and would then discard the first sheet so that it would never be used again. The receiver would decrypt it also using the first sheet, and then discard the sheet. This cipher, the *one-time pad*, was invented by Joseph Mauborgne for the US Army at the end of the First World War, and is known to be unbreakable by anyone not in possession of the one-time pad. Its major limitation is the necessity for producing and securely distributing the pad.

In fact the process of inventing better ciphers (by those trying to send and receive secure messages) and devising ways of breaking them (by their enemies) is a game people have played for millennia.

Methods and machines

Given that the processes of encryption and decryption are normally based on well-defined rules, it's a little surprising that the use of mechanical aids was relatively slow to get going. Simple substitution ciphers require no more than a two-row table: plain-text letters on the top row and substitutes on the bottom. The Vigenère cipher requires a square table, with each of the 26 possible alphabet shifts on its own row. Even the one-time pad is essentially paper-based.

However, it is also possible to make a simple mechanical device to help with either the simple substitution or Vigenère-style encryption and decryption, in the form of a pair of disks, one inside the other. The letters of the alphabet are written around the edge of each disk, and the inner disk is rotated relative to the outer disk to set up a single substitution table. If it is further rotated during encryption, a Vigenère-style cipher is produced.

Such a disk was invented by Leon Alberti in the fifteenth century, and similar devices were in use for a long time, including during the American Civil War. Perhaps surprisingly, it was not until the twentieth century that the use of machinery for encryption and decryption advanced much further. However, the application of a complex cipher system really does suggest or even demand machinery: the more complex the rules to be applied, the more important it is to delegate their operation to a machine, which might be expected not to make mistakes.

Mechanisation of encryption and decryption did not really take off until the invention of the Enigma machine. The German military famously used Enigma as their preferred cipher device during the Second World War, both for encryption and decryption, with daily changing keys; and the British, equally famously, had at Bletchley Park an establishment devoted to reading German cipher messages, which did in fact repeatedly and successfully break these daily ciphers.

Enigma

Enigma generates a letter-by-letter substitution of the clear message, but the substitution table effectively changes with every letter. But unlike the original keyword-based Vigenère system, the table does not repeat itself every few letters. It is more comparable to the one-time pad.

It is a fascinating machine in its own right. Developed in 1918 by Arthur Scherbius, it looks very much like a typewriter—in fact the keyboard is closely based on Sholes' keyboard described in Chapter 5 (which by 1918 was well established as the standard form of keyboard for typewriters). But instead of paper, the back of the machine has a replica of the keyboard in a lampboard, an arrangement of lettered disks each with a lamp behind it—see Figure 19. (You might note in passing that this keyboard differs a little from the Sholes typewriter keyboard (Chapter 5), although obviously derived from it. In particular, the offsets differ—having only three rows, the offsets are one-third of the key width. Experienced touch-typists would have noticed this!)

The letter L is pressed, and the D lamp is on.

Figure 19: Enigma machine
https://commons.wikimedia.org/wiki/File:
Enigma_Machine_A16672_open,_letter_L_pressed.agr.jpg
CC BY-SA 4.0.

The clear message is typed in, as it might be on a typewriter, but at each keystroke, instead of printing, one of these lamps is illuminated, indicating a new letter—the cipher code to be used for the letter just typed in. The illuminated letter then has to be recorded somehow—written down or typed or transmitted directly.

The mechanisms that allow the continually-changing table of substitutions are several and ingenious, and I will not attempt to describe them here. They depended on initial settings, which were changed daily; once into a message, the settings were changed automatically by the process of typing the message. That is, every keystroke resulted not only in the coding of one letter of the message, but also in re-arranging the table of correspondences for the next keystroke.

The resulting cipher was extremely complex and difficult to break, but the complexity arose not so much from complex rules, as from a combination of many applications of simple rules. This is exactly the province of the machinery of the time, and it is no surprise that encryption and decryption should have succumbed to some such form of mechanisation, not long after the typewriter and the comptometer.

Breaking Enigma

As I have indicated, code-breaking is a different order of task altogether. Almost inevitably, given a series of cipher messages, breaking the cipher system involves a combination of knowledge or guesswork as to the mechanisms involved in the cipher (or rather the rules which they mechanise), knowledge or guesswork about some key settings, and trial and error. There is a very strong sense in which code-breaking is an art-form. Like any art form it has its supporting technology (both in the form of machinery and in the form of know-how, methods and ways of doing things, sets of rules that may be applied), but it needs inspiration as well. This is certainly not true of encryption or decryption. It could be said to be true of devising new cipher systems—and indeed there are a couple of late-twentieth-century inventions here that look truly inspired—but as a hothouse for developing new ways of thinking, it is hard to beat Bletchley Park.

Bletchley Park was the Second World War UK Government establishment in charge of attempts to read any intercepted cipher enemy messages. Many messages to or from military units of all kinds, of the German or other Axis powers, were intercepted and sent to Bletchley Park. And for much of the war many of these messages were successfully decrypted. There was always a challenge at the beginning of the day, because the keys were changed each day and the new key had to be discovered from some of the early messages intercepted. Then for some longer intervals, weeks or months, a particular cipher might become unreadable because of a change in some part of the encryption procedure by the Germans—until the Bletchley Park people had discovered how to deal with this new variant.

Post-war cryptography

In the subsequent history of cryptography, following the end of the Second World War, the computer has loomed large. Most modern cipher systems are computer-based, in the sense that computer programs are used for encryption and decryption as well as by the code-breakers. In fact most systems make use of the fact that a message in a computer (necessarily in one of the binary codes discussed in Chapter 5) looks very much like a number, to which arithmetic operations can be applied. Of course it isn't really a number, but with certain safeguards we can pretend that it is. We can encrypt by applying arithmetic operations to it such as addition and multiplication; decryption then means reversing these operations.

One of the great discoveries of cryptography in this period is the principle of asymmetry. This is based on the fact that some arithmetic operations are easy to perform in one direction, but much harder in the other (it's easy to multiply two large prime numbers; it's much harder to factor the product and rediscover the original two primes). The resulting cipher system is known as Public Key Cryptography. It allows the person who wants to receive a message in cipher to make public an encryption key; anyone who wants to send him/her a message can use this key to encrypt it. However, the decryption key is different. Only the recipient of the message knows this decryption key—it need never be made available to anyone else. In principle, this makes for a much more secure setup—in almost all previous cipher systems, sender and recipient would have to share a key, and the necessity for sharing is a major source of insecurity.

Bletchley Park and its legacy

Despite the fact that cryptography really entered the machine age only after the First World War, the challenge of cryptanalysis and code-breaking must really be credited with kick-starting the IT revolution of the second half of the twentieth century. In the end, we did not invent computers in order to control machinery, as Jacquard might have done; we did not invent computers in order to do repetitive numerical calculation, as Babbage tried to do. We did not invent them to analyse censuses; nor to organise our accounts or do payroll; nor to do weather forecasting; nor to do word processing; nor to facilitate telecommunications; nor to play our music or look after our photographs—though they are very useful for all of these things and more. We invented computers in order to break codes.

The operation of Bletchley Park depended very heavily on people: collecting, transcribing, analysing the intercepted cipher messages. Initially, all analysis was entirely by people, using essentially pencil and paper, and human effort remained central to the code-breaking task. However, early in the war the great Alan Turing designed a machine called a *bombe*, which greatly helped in eliminating many possible initial settings (given a *crib*, a human guess as to the plaintext version of a particular section of the cipher text). This invention allowed Bletchley Park, for much of the war, to discover the day's new key settings early in the day, enabling the decryption of any further messages that day as soon as they were received.

Later in the war, the Bletchley Park effort had serious difficulties with another German system, the Lorentz cipher. This was similar to Enigma but more complex, and it typically took weeks to break one day's messages. Max Newman, another Bletchley Park mathematician, started developing plans for a new machine that would be much more adaptable than the bombe—in fact, it was what we now describe as *programmable*. This was much more difficult to build than the bombe, but eventually in late 1943 the engineer Tommy Flowers designed and constructed a working version, using thermionic valves (as used in early radios). It was called the Colossus, and with its help, the keys for Lorentz-ciphered messages could be discovered quickly.

Colossus was the clear forerunner of the modern computer. It was electronic, digital, and in some sense programmable, and used many of the ideas and principles and methods that a modern computer scientist would regard as essentially those of a computer.

An act of vandalism

Then, at the end of the war, the entirety of what had been the Bletchley Park operation was eliminated. Winston Churchill, who had been the chief backer of Bletchley Park, ensuring funding for it against opposition from some quarters, demanded that all evidence of the UK's cryptographic abilities should be utterly erased. Not only was Colossus itself destroyed, but all the blueprints for it were burnt. All Bletchley staff were required to keep silent about anything at all that went on there.

Despite my heading, *vandalism* is a poor word to describe Churchill's action. It was a 2000-year throwback to the first emperor of China, in the second century BCE—burning the library, in order to suppress the subversive knowledge held therein.

The next phase

But it's hard to kill an idea like that. In the world of the 1940s, outside Bletchley Park, some of the necessary ideas were already coming together. A project between IBM and Harvard University, masterminded by Howard Aiken, developed the Harvard Mark 1, a giant programmable calculator with many computer-like features, which first ran in 1943. The destruction of Bletchley Park left behind, in addition to the handful of eccentrics who believed in the possibility of building a computer, another handful who had actually seen one in operation. Within a year or two immediately following the war, academics in the UK (at Manchester and Cambridge) and in the US (in Pennsylvania and elsewhere) started building computers. Within a very few years, the computer age had taken off.

But that's another story.

Epilogue

We have seen a skein of different ideas, developing over the course of human history, interacting with and feeding off one another, brokered by people with a wide variety of different motivations. We have seen the notion of *data* emerge gradually and gradually absorb many other concepts. Information, which might be seen as an abstraction like *matter* or *energy*, is in some sense "carried" by data, or perhaps may be extracted from it. Numbers are data, text is data, pictures are data, music is data. But that's just the beginning—now everything we do, every interaction we have with any part of the world around us, is data.

Of course this is all absurd. Music (just to take one example) is a human experience, or rather a whole raft of human experiences, and to regard it as data is to ignore or put aside both the nature and the validity of the experience, whether of composing or of performing or of listening. Nevertheless, it is convenient to pretend that music is data, because there is so much we can do with it on the back of that pretence. Not only can we record, store, retrieve, transmit, broadcast music-as-data, we can also make use of any number of digital tools (as well as the slightly older analogue electronic ones) as part of the process of creation, in both composition and performance.

In the twenty-first century, data, data processing and manipulation, and all the raft of technologies around data, are central to how we see the world. In these days of data protection and privacy, of laws and regulations around this domain, of data mining, of data theft, of people and organisations who relentlessly collect data about us and who manipulate us by manipulating our data, and so on—in these days, it is hard to re-imagine the world as it was before the notion of data took hold. The digital computer—together with all the other information and communication technologies—is of course at the core of this data-centred world. Which is why it is tempting to speak of the invention of the computer having ushered in a

    https://doi.org/10.11647/OBP.0225.13

revolution.

So, was it a revolution? Did the arrival of computers result in an over-throw of the existing order of things and its replacement by something fundamentally new?

Certainly, the effect on our lives of the developments in the domain of the information and communication technologies, subsequent to and at least to some extent consequent upon the invention of computers in the 1940s, has been immense, arguably revolutionary. The world of email, the internet, online shopping, online management of bank accounts, mobile phones doubling as cameras, digital radio and television, downloaded recorded sound and films, satellite navigation, ebooks, Google, Wikipedia, social media—all this would have seemed utterly extraordinary, something in the realm of fantasy, to my parents at the time I was born.

Nevertheless, the existing order is seldom so easily cast aside. What this book has demonstrated, I hope, is the extraordinary amount of stuff—of knowledge, understanding, invention, ways of thinking and doing, ideas, methods and techniques—we have brought with us over this journey. In many significant ways, the IT world not only draws on the past, but is rooted in it. This past is not just (though it very much includes) the couple of centuries following the industrial revolution, but goes way back—to the Renaissance, to the invention of printing, to the ninth-century Arabic and seventh-century Hindu mathematicians, to the Roman empire, to the Greeks and the Phoenicians, to the invention of writing itself.

Bibliography

What follows is a short list of books and essays (and one film) that have inspired me and that I have mentioned in this book for one reason or another. But I must give one reference pride of place as my go-to source of first and often only resort:

Wikipedia: Many articles, by many authors.

—

Jim Al-Khalili, *Pathfinders: The Golden Age of Arabic Science* (Penguin, 2012).

Isaac Asimov, *Foundaton Trilogy* (Gnome Press, 1951).

Antonio Badia, *The Information Manifold: Why Computers Can't Solve Algorithmic Bias and Fake News* (MIT Press, 2019). https://doi.org/10.7551/mitpress/12061.001.0001

The Venerable Bede, *On the Reckoning of Time* (c. 723).

Ray Bradbury, *Fahrenheit 451* (Ballantine, 1953).

Vera Britain, *Testament of Youth* (Victor Gollancz, 1933).

Lewis Carroll, *Through the Looking-Glass, and What Alice Found There* (1871).

James Essinger, *Jacquard's Web: How a Hand-Loom Led to the Birth of the Information Age* (Oxford University Press, 2004).

Luciano Floridi, *Information: A Very Short Introduction* (Oxford University Press, 2010). https://doi.org/10.1093/actrade/9780199551378.001.0001

James Gleick, *The Information: A History, a Theory, a Flood* (Fourth Estate, 2011).

Stephen Jay Gould, 'The Panda's Thumb of Technology', in *Bully for Brontosaurus: Reflections in Natural History* (Penguin, 1992).

GPO Film Unit, *Night Mail* (1936).

Robert Kaplan, *The Nothing That Is: A Natural History of Zero* (Allen Lane, 1999).

John Man, *Alphabeta: How 26 Letters Shaped the Western World* (Headline, 2000).

John Man, *The Gutenberg Revolution: The story of a genius and an invention that changed the world* (Review, 2002).

Cyril Northcote Parkinson, *Parkinson's Law: The Pursuit of Progress* (Houghton Mifflin, 1957).

Steven Pinker, *The Language Instinct: The New Science of Language and Mind* (Harper, 1994).

Andrew Robinson, *The Story of Writing: Alphabets, Hieroglyphs and Pictograms* (Thames & Hudson, 1995; 2nd ed 2007).

David Rothenberg, *Hand's End: Technology and the Limits of Nature* (University of California Press, 1995).

Paul Saenger, *Space Between Words: The Origins of Silent Reading* (Stanford University Press, 2000).

Claude Shannon, 'A Mathematical Theory of Communication' *The Bell System Technical Journal* XXVII, No. 3 (July 1948), pp. 379–423.

Adam Silverstein, *Postal Systems in the Pre-Modern Islamic World* (Cambridge University Press, 2007).

Simon Singh, *The Code Book: The Science of Secrecy from Ancient Egypt to Quantum Cryptography* (Fourth Estate, 1999).

Tom Standage, *The Victorian Internet: The Remarkable Story of the Telegraph and the Nineteenth Century's On-Line Pioneers* (Walker, 1998).

JoAnne Yates, 'Co-evolution of Information-Processing Technology and Use: Interaction between the Life Insurance and Tabulating Industries' *The Business History Review* 67, No. 1 (Spring, 1993), pp. 1–51.

List of illustrations

Index of topics

Index of names

About the team

Alessandra Tosi was the managing editor for this book.

Lucy Barnes performed the copy-editing and proofreading.

Anna Gatti designed the cover using InDesign. The cover was produced in InDesign using Fontin (titles) and Calibri (text body) fonts.

The author typeset the book in LaTeX. The text font is Tex Gyre Pagella. The author generated the PDF from which the paperback and hardback were produced, as well as an ebook version.

Luca Baffa created the remaining versions (epub, mobi, HTML) from the ebook version. The conversion is performed with open source software freely available on our GitHub page (https://github.com/OpenBookPublishers).

This book need not end here…

Share

All our books — including the one you have just read — are free to access online so that students, researchers and members of the public who can't afford a printed edition will have access to the same ideas. This title will be accessed online by hundreds of readers each month across the globe: why not share the link so that someone you know is one of them?

This book and additional content is available at:

https://doi.org/10.11647/OBP.0225

Customise

Personalise your copy of this book or design new books using OBP and third- party material. Take chapters or whole books from our published list and make a special edition, a new anthology or an illuminating coursepack. Each customised edition will be produced as a paperback and a downloadable PDF.

Find out more at:

https://www.openbookpublishers.com/section/59/1

Like Open Book Publishers

Follow @OpenBookPublish

Read more at the Open Book Publishers BLOG

You may also be interested in:

Engaging Researchers with Data Management
The Cookbook

by Connie Clare, Maria Cruz, Elli Papadopoulou, James Savage, Marta Teperek, Yan Wang, Iza Witkowska, and Joanne Yeomans

https://doi.org/10.11647/OBP.0185

Writing and Publishing Scientific Papers
A Primer for the Non-English Speaker

by Gabor Lovei

https://doi.org/10.11647/OBP.0235

The Essence of Mathematics Through Elementary Problems

by Alexandre Borovik and Tony Gardiner

https://doi.org/10.11647/OBP.0168

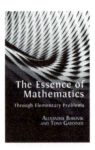

www.ingramcontent.com/pod-product-compliance
Lightning Source LLC
LaVergne TN
LVHW061955050326
832903LV00036B/4829